BANK INTERIOR

银行室内空间

高迪国际出版有限公司 编
薛昱 钱律伟 罗小波 王高思 洪巧红 译

大连理工大学出版社
Dalian University of Technology Press

图书在版编目(CIP)数据

银行室内空间：汉英对照/高迪国际出版有限公司编；薛昱等译. — 大连：大连理工大学出版社，2012.3

ISBN 978-7-5611-6740-3

Ⅰ.①银… Ⅱ.①高… ②薛… Ⅲ.①银行—室内装饰设计—图集 Ⅳ.①TU247.1-64

中国版本图书馆CIP数据核字（2012）第016245号

出版发行：大连理工大学出版社
　　　　　（地址：大连市软件园路80号　邮编：116023）
印　　刷：利丰雅高印刷（深圳）有限公司
幅面尺寸：240mm×320mm
印　　张：24
插　　页：4
出版时间：2012年3月第1版
印刷时间：2012年3月第1次印刷
责任编辑：刘　蓉
责任校对：李　雪
封面设计：屈舒丽

ISBN 978-7-5611-6740-3
定　　价：380.00元

电　话：0411-84708842
传　真：0411-84701466
邮　购：0411-84703636
E-mail：designbooks_dutp@yahoo.cn
URL：http://www.dutp.cn

如有质量问题请联系出版中心：（0411）84709246　84709043

PREFACE_A
序言_A

Over the years the bank branches resembled post offices in the way they operated. A bank was a place visited rather frequently and out of a necessity, the clients were treated in a much more formal and bureaucratic manner. When money lost its material tangibility and thus became more accessible with the introduction of credit cards, when many bank services became available outside the branch by phone and Internet, the relationship between the client and the financial institution managing his wallet had to change as well. The bank branch had to adapt to new functions and tasks that appeared together with technological developments.

What is peculiar, the dynamics of the change in this sector seems to be the strongest in developing countries. The lack of a long-standing tradition of commercial banking made technological revolution go hand in hand with innovation in thinking about how the banking institution should function to gain customers' trust. A niche in the market had to be filled – many financial institutions were launched proposing bold design solutions, in comparison to the overall conservatism of a bank image in the developed countries.

Banks today are comfortable and attractive places, with a distinctive interior design which, together with other elements of brand identification, creates a consistent message and manifests "the personality" of the institution. Coherence and authenticity of this message has a significant impact on the bank's market success. It also proved to be that non-obviousness and experiment, innovative and often spectacular solutions have become an effective method to attract the attention of customers.

The design of a modern banking facility is a challenge to create a place both functional and of an expressive image. Previously ignored or neglected, the way the branch looks and works has become a very important factor in competitiveness. The banking facility as part of the human environment should be a place made to achieve a certain effect. Through a conglomerate of impressions in well-designed space a deeper and more direct relation between the bank and its client can be developed. The bank's interior is a statement that can influence the customer in a way unavailable to other forms of communication, and has become one of the most important tools of gaining the hearts and wallets of clients.

Robert Majkut

多年以来，银行分行的经营方式与邮局类似。银行是人们经常去的一个地方，也是人们生活的一部分，工作人员对待客户的方式显得更加正式和官方。当金钱变得无形，信用卡出现时，钱就更容易得到了。当许多银行服务不需要去银行就可以通过电话和互联网得到时，客户与管理其资产的金融机构之间的关系也就不得不改变了。银行分行只得适应那些伴随技术发展的新功能和任务。

令人诧异的是，在发展中国家，银行部门的改变显得更加明显。由于缺乏商业银行那种持久的传统，科技革命与革新密切相关，而这种革新在于金融机构该如何运转以获得顾客的依赖。市场的空缺不得不填满——那些提出大胆的设计解决方案的金融机构不断建立起来，这与发达国家中整体保守的银行形象恰恰相反。

今天的银行是舒适且富有吸引力的地方，它的内部设计与众不同，再加上可以展示银行身份的其他元素，传达了一致的信息，同时也显示了这个机构的"个性特点"。一致可靠的信息对于银行能否在市场上取得成功具有重要意义。它同时还证明了创造性的、具有实验意义的、革新且壮观的解决方案已经成为一种可以吸引客户的有效方法。

设计一家现代的金融机构是一个挑战，这个挑战即建造一个功能齐全，又令人印象深刻的地方。先前被忽略的东西，即分行的外观和工作方式在市场竞争中有着非常重要的影响。作为人类环境之一的金融机构应该是一个可以产生某些影响的地方。精心设计的空间给人很多不同的印象，在这种影响下，银行与客户就可以建立一种更加深入而直接的联系。银行的室内设计是一个很好的陈述，它可以以一种非口头交流的方式来影响顾客，同时它也变成了获得客户信赖和钱包的重要的工具之一。

Robert Majkut

PREFACE_B
序言_B

Designing the Interiors of Banks has become a fiercely contested target market. Design firms are being stretched by the large Banking groups to be more and more creative as this traditionally commercial design field increasingly blurs between the retail, corporate and hospitality sector. As the demands on Designers intensifies and some of the emerging more agile firms respond to these challenges there is quickly becoming a new Specialist area, and that area is specifically Bank Interiors.

Gone are the days of simply repeating the same old concept. There is steep competition between the many differing Banking Groups in every country, every City and every town. Each area within each of these locations has a different commercial demographic; furthermore the commercial environment within which all of these Banks and their many branch outlets operate is continually changing. Each Bank has its own Brand or series of Brands and all of these brands are required to respond to the commercial demand specific to the location of that branch. Most of the large Banking Groups offer services across large areas of the world, all of the various locations have unique security requirements and many have strict planning rules and regulations. All of these constraints are growing, not shrinking, as the competition becomes ever more sophisticated and aggressive.

There are specific departments within large Banks who gather the necessary information for strategists to determine market forces. "Local" Architects within each Country and Engineers provide data on regulations. Also the ongoing brand standards and emerging market brand strategies have a major role in the design direction. The challenge to Interior Design firms is to digest all of the available information and to provide a solution which adds value to the existing brand, appeals to the correct customer base, and one that enhances the customer experience. The design has to be specific to the market force for that project. It is for these reasons that Banks can no longer remain competitive by simply "Rolling Out" the same old concept. And it is for these reasons that this new Design market has emerged and is being developed into a more diverse dynamic.

It is interesting to see in this publication examples of new sectors within Retail Banking which have emerged in response to an ever broadening client base. One such area is a more private wealth management service for High Net Worth Individuals with levels of liquidity typically at around 2m. This sector has started to provide environments which are more Hospitality and less Commercial in approach to the Interior design solution. As this market grows we are seeing more exclusive groups with ever higher liquidity entry levels. These individuals and groups expect exceptional service in environments which are less related to Bank Interiors and more to that of a private club or Hotel Lounge.

I have been working in the Hotel and Restaurant Design markets in both London and Asia for 25 years and it is interesting that the "experience" oriented environment of these markets has started to take root in Banking Interior Design. It throws new light onto the subject and gives the Client fresh eyes. The evolution of this process has been showcased in this book, the design is of course subjective, but what is indisputable is that this selection of new Bank Interiors illustrates that Bank Design is now a new and exciting specialty field.

Patrick Waring
——Silverfox Studios

　　银行大厦的室内设计已经成为一个竞争激烈的目标市场。设计公司为银行集团设计的现象越来越普遍,因为这个传统的商业型的设计领域越来越难定义,它的功能性质游弋于零售业、企业和酒店业之间。但由于银行集团对银行大厦设计方面的要求越来越高,越来越多能够灵活应对、满足客户要求的专业化设计公司的纷纷涌现,使得银行大厦的建筑设计这块领域越来越专业化,尤其是银行大厦的室内设计。

　　那些只是简单地重复着旧理念的时代已不复存在。在每个国家、每个城市、甚至是每个乡镇,银行与银行之间都存在着十分激烈的竞争。每个区域都有其独特的商业人口;此外,在每个特定区域的银行的商业环境以及它们分行服务点的操作方式也一直在不断地变化着。每家银行都有自己的品牌或品牌系列,而这些银行必须回应针对该银行分行所在区域的商业需求。多数大型银行集团的业务都是遍布全球的,所有的分支点都有其独特的安全需求,其中有些还有严格的规划法规和规章制度。由于竞争变得更加激烈和复杂,使得所有这些限制因素没有任何的放宽倾向,反而日益加剧。

　　大型的银行集团都有特定的部门收集必要的信息以制定本土化的市场战略;而这些设计师和工程师在其特定的国家为银行提供数据管理。同时,现有的品牌标准和和新兴市场的品牌战略在设计方面起着十分重要的作用。室内设计公司要做的是提取所有有用的信息,并且提出一个解决方案,融会贯通,使其能够成为现有的品牌价值的一部分,吸引客户群,同时提升客户在银行所享有的服务体验。银行大厦的设计必须本土化。正是因为这些原因,银行不能再以简单地"复制"相同的、过时的理念来保持竞争力。也正是因为这些原因,这个新兴的设计市场才会形成,并且发展得更活跃、多样。

　　在这本书内所提及的银行大厦案例中为了响应不断扩大的客户群而成立的一些新的银行业务部门都十分有趣。其中就有一个专为流动资金至少约200万的高资产净值人士而打造的私人财富管理服务。这个部门的最初的意图是通过重塑室内设计给客户营造一个多一份热情友好、少一份商业气息的服务环境。随着这方面市场的开发,我们现在可以看到很多更高流动资金限额的、更加专业化的私人财富管理服务。这些个人或团体期望能够通过提供银行大厦的室内设计环境的卓越服务,将银行打造成为一个不仅仅只是提供银行业务的场所,同时还是私人俱乐部或是私人商务会所。

　　我从事于亚洲和伦敦市场的酒店和餐饮场所的室内设计已有25年了,我觉得这种热衷于提高环境方面的客户服务体验的概念渐渐融入到银行大厦的室内设计中是十分有趣的。这种新颖的设计元素会给银行客户们带来眼前一亮的感觉。这种演变过程在本书中有详细的介绍。当然设计是带有主观色彩的,但不容置疑的是,越来越多的倾向于新型银行大厦室内设计的选择很好地证明了现在的银行建筑设计是一个新的令人振奋的专业领域。

Patrick Waring
——Silverfox Studios

PREFACE_C
序言_C

onespace
GLOBAL BANKING INTERIORS

Few interior design typologies have evolved as dramatically in the past 15 years as have banking interiors. Moreover, just as there is no single, homogenous form of "restaurant", there is no single "banking" environment either. A very wide range of banking operations, regulatory controls and customer relationships exist, which require design and planning distinction from one another in profound, yet sometimes subtle, ways.

Recognising this, my partner and I established One Space several years ago specifically to focus on the rapidly evolving needs of the financial services sector. Since that time, we continue to witness an extraordinary shift in market perceptions and management priorities that naturally tend to respond to the global economic situation.

One particularly significant phenomenon, globally, has been the heightened focus on emerging markets. For our clients, this has meant that our interior design solutions must be more responsive, adaptable and agile than their predecessors. Gone are the days when bank executives sat behind imposing mahogany-panelled doors barking directives to their juniors. By contrast today, extensive peer-to-peer cooperation across the bank's businesses and a more unified approach to its markets necessitates greater executive interaction, staff engagement, and collaborative teamwork.

Uniquely, our practice integrates architecture, interior design and technology, because these market-driven transformations (demanded by the business of banking, and brought to life by our interior design and planning) are enabled only by the technology advances that underpin them. For example, many of our global clients are travelling less – partly in response to climate change and corporate responsibility goals – whilst cooperating more closely across geographies, so that virtual meeting venues and installations, along with collaboration tools, have become a priority.

Internet banking and self-service transactions have also made a mark on the design of banks, particularly in the retail banking sector. In fact, these on-going changes, accompanied by a more tech-savvy generation of consumers, increased market segmentation and changing demographics of wealth and spending, have driven retail and commercial banks to wholly re-examine their value-adding proposition.

Along with emerging markets comes rising affluence in many parts of the world, and banks have needed to respond to the changing demographic of high-net-worth clients. The rise of private wealth management as a global bank offering presents a unique set of design challenges. Universally, these wealthy clients have an expectation of privilege, exclusivity and stature, but the manner in which the interior setting attends appropriately to this narrow but valuable customer base is directly informed by the cultures and generations from which they come. Our multinational bank clients now must reconcile their globally-recognized brand identities with their customers' local cultural expectations, and our designs become a crucial instrument in conveying deftly those important first impressions.

An overview of bank interiors globally would not be complete without mention of governance. Inevitably, the events of recent years implore banks to put increasing emphasis on demonstrating value for money. There is heightened due diligence surrounding decision-making on major capital spending, as well as making notions of transparency physically manifest. But, as in all design endeavours, aspirations are balanced by constraints; and design constraints, in this context, are treated not as limitations, but, on the contrary, as a creative stimulus.

Greg Pearce

很少的室内设计类型学能像银行室内设计学一样,在过去的15年内发生了翻天覆地的变化。此外,正像没有独立、单一形式的"餐厅"一样,同样也不会有独立的银行环境。在很大范围内、银行的运行体系中都会存在法规控制和客户关系,这就需要银行在设计和规划方面,有一种深刻或者微妙的区别。

因此,我和我的伙伴在几年前创建了 One Space,专门为满足金融服务部门快速的发展需求。从那以后,我们继续见证了在全球经济形式的引导下,市场认知和管理优先级的惊人转变。

一个尤其重要的现象已经形成,那就是全世界都更加关注新兴市场。对于我们的客户来说,这就意味着我们的室内设计方案必须比他们的前任合作商更具敏感性、适应性和灵活性。那些银行主管们坐在红木格子门后面对他们的晚辈发号施令的年代已经过去。相比之下的今天,贯穿银行业务的广泛、平等的合作以及一种进入市场更为统一的途径,促使更为强大的行政互动、全员参与和合作精神成为必须。

因为这些市场驱动的转变(受需求于银行的业务,通过我们的内部设计和规划引入生活)是由支撑它们的技术的进步来实现的,所以我们采用了独特的方式——将建筑学、室内设计和技术一体化。比如,我们许多国际性的客户较少出去旅游——其部分原因是由于不适应环境的变化和公司责任的目标限制;同时,他们又要加强跨越地域的合作。因此,虚拟会议室场地、设施以及协作工具便获得了优先权。

在银行的设计中,尤其在零售银行业、网上银行和自助交易已经开始崭露头角。事实上,这些正在进行的增加了市场划分,改变了财富和消费的人口统计资料,伴随着一代更精通科技的消费者,已经驱使零售和商业银行对它们的增值提案进行彻底的再检验。

随着新兴市场在世界大部分地区日益富裕,银行必须对高净值客户不断变化的人口统计做出回应。作为全球银行新股,私人财富管理的崛起展示了一项独一无二的设计挑战。一般来说,这些有钱的客户期望着一种特权性、专有性和高度性;然而内部设置适当参与到这类稀少而又珍贵的客户群的方式,是通过他们的源头文化和世代直接传递的。如今,我们的跨国银行客户必须使全球认可的品牌形象和他们的当地文化期望相一致。因此,我们的设计就成了熟练传达这些重要第一印象的决定性工具。

不形成一种管理,全球的银行室内设计就不会有完整的概观。不可避免的是,近年来的各种项目好似在哀求银行应该更重视对金钱价值的展示。在主要资本开支和透明的事物清单观念形成方面,有一种更高要求的审慎调查的环境决策。但是,就像在所有的艰苦设计中,有多少期望,就会有多少约束。这种设计约束,并不是对你的思维进行限制,而是激发你的创造性。

Greg Pearce

PREFACE_D
序言_D

NEW METAPHORS FOR THE BANK

Credit institutes have always entrusted their image to the classic iconography of safety: the fairly anonymous expression of reliability represented by materials such as marble, metal, and wood, and formal solutions designed to transmit the impregnability of the place where the money is kept. The creation of a deep environmental fracture between the user and the service provider is the most direct consequence of this approach.

The image of the bank has changed in many ways in recent years, and been redirected above all in the direction of a better relationship between the bank and the public. The expressions "calm-familiarity", "the bank must never be ostentatious", and even "transparency and reliability" have all become passwords in the design of the layout to the extent that the concept of "user-friendly" has been overworked into a cliché that must now be overcome. With nearly twenty years of experience in the sector, our studio has always attempted to assume an experimental attitude that restores strong symbolism and subtle irony to the image without neglecting comfort or hindering the execution of daily bank activity through the latest methods right from the start.

Working with smaller banks, we have created many interlocking pieces that have all played parts in building the total image of each institute.

Which message must be conveyed to the customer? "Joy!"... and so "joy" must be reflected in a service filled with joy. If the service is then rendered with joy, we can make one branch different from the next, and anyone who steps inside will remember it forever.

In the end, a bank is always a sales outlet, even if the goods are not arranged on shelves. Why not represent the bank as a stage set with quality? A place where communication prevails over function, a place where an event is characterized also by its place of occurrence and one where even our emotions have crystallized?

For example, we once had to create a branch in a small town's historical center where an abandoned water mill stood. We were able to imagine the potential the moment we saw the site: the wooden flour bin was still there, together with an old hayloft and other old agricultural tools. In our renovation of the building, we left the flour bin in place as if it were an enormous sculpture, converted the hayloft into office space, and put the old implements on display in glass cases. The result was a highly personalized bank branch, one that was also deeply integrated in its surroundings and the local community: "Water Mill with Bank".

On another occasion in Montelupo Fiorentino near Firenze, potters and ceramic artists provided the context. Here, we set up a special area inside the bank to be dedicated to exhibitions by different artists. The rest of the space was distinguished by large decorative pottery resembling large still life works looking down from the walls above. The overall effect was "Still Life with Bank". I think banks should be designed in a way that makes users feel good about coming inside and transforms bank operations into pleasant rituals. The design should create advantageous conditions for both the bank and the customer.

Scenery developed using metaphors evoking distant places can also be created today for ambience that has much less to do with the workplace than in the traditional sense. A simple closet can become a sculpture that also serves the purpose of storage. Money is still money, but the surroundings can change. It's nice to think that at least here, and now, the container can become more human, more romantic. And maybe tomorrow, the money will follow.

Massimo Mariani Studio

　　信用机构总是以安全性能极高的样子示人：使用大理石、金属和木头等建材来体现其坚不可摧似乎成了一种不成文的规定，而这种建造带来的最直接的结果就是信用机构和客户间出现了一道坚固的隔阂。

　　近些年来，银行的形象在许多方面都发生了变化，而最重要的变化莫过于它在银行和客户关系改善中扮演的角色的转换。诸如"冷静的熟悉"、"银行不得炫耀"及"透明度和可信度"等词都成了银行设计建造时的关键词，甚至从某种角度来说，像"用户友好型"等词已成了需要摒弃的陈词滥调。经过近二十年在这一领域的探索，我们的工作室勇于尝试，在重建强烈的象征主义和微妙的反讽的同时，不忘使用最新的技术将银行的日常事务规整得舒适和畅通。

　　在与一些小银行的合作中，我们创建了许多互相关联的个体建筑，它们最后都能有机地合在一起，表达出一个银行的整体形象。

　　我们最应该传递给客户什么呢？"愉悦！"。因此，"愉悦"必须在充满愉悦的服务中传递给客户。如果服务能在令人愉悦的环境下进行，那么我们的分行就可以变得与众不同，来过的人们都将难以忘怀。

　　最后，银行也是一种销售途径，尽管它的商品不是直接陈列在架子上。为什么不将银行布置得漂亮上乘呢？比如，把它建造成一个交流比其本身功能更重要的地方，一个发生地点能影响事情的地方，一个我们的情绪能被量化的地方？

　　比如，我们曾将一个分行造在了一个小镇的历史中心处，那里还留有一个老旧的水磨坊。我们第一眼看到这个地方就觉得这个地方潜力极大：面粉桶、干草棚和其他古老的农用工具都还在。在我们翻修时，把面粉桶留在了原地，看起来就好像是一个巨大的雕塑；我们还把干草棚设到了办公室里面，再将古老的农用工具放在玻璃橱窗里做展示。结果这个分行成了一个极具个性的银行，它与周围的事物和当地的环境完美地融合在了一起，美其名曰"水磨坊银行"。

　　另一个例子是弗罗伦萨附近的小镇 Montelupo Fiorentino，该镇上的陶工和陶器艺术家们给了我们灵感。我们在银行内设置了一块专门的区域用来展示艺术家们的陶器艺术品，其他地方则摆上巨大的装饰性陶器，从上往下看，就像是巨大的静物画一样。银行就成了一个"静物画银行"。我认为成功的银行设计应该让顾客感觉舒适、愿意进门，而银行的例行公事也变成令人愉快。银行的设计应该为银行本身和顾客创造良好的条件。

　　通过模拟营造远处环境发展而来的景物如今可以在周围环境中创造出来了，相对于传统的建造方式来说，这与实际工作地的联系性要小的多。我们可以将一个简单的壁橱变成一个好看的雕塑，同时又保留住它用来储藏的实用价值。钱还是钱，但是放钱的地方可以发生改变。至少我们看到储钱库可以变得更加人性化，更加浪漫。也许明天，钱也会变成这样。

Massimo Mariani Studio

CONTENTS

CONTENTS
目录

14 ING BANK

24 CORPORATE DEPARTMENT OF ING BANK ŚLASKI

34 CHEBANCA!

46 NATIONAL BANK OF OMAN

54 BARWA BANK

62 DEUTSCHE BANK

70 HSBC SHANGHAI

80 OPEN LOUNGE – RAIFFEISEN BANK

90 LLOYDS BANKING GROUP

98 STANDARD CHARTERED BANK

108 BANK SAUDARA PRIORITY BANKING

116 BCC BANK

126 NEW OFFICE FOR RABOBANK NETHERLAND

134 BNP PARIBAS BANK

144 ABU DHABI ISLAMIC BANK

152 BANCO DE CRÉDITO DEL PERÚ

162 BANCOLOMBIA

170 I-BANK

178 AIR BANK

186 EXTRABANCA

192 DEUTSCHE BANK, THAILAND

198 BANK IN DONORATICO

204 MIDFIRST BANKING

214 NATWEST

218 UOB BANK

230 ONECALIFORNIA BANK

236 BANK IN MONTELUPO FIORENTINO

244 WORLD BANK

252 BANK OF MOSCOW

262 ALIOR BANK'S PRIVATE BANKING CENTER

272 NOBLE BANK

280 HSBC

286 CITIGOLD SELECT

296 INDEPENDENCE BANK HQ

304 INDUSTRIAL AND COMMERCIAL BANK OF CHINA

310 JULIUS BAER PRIVATE BANK

316 ZÜRCHER KANTONALBANK

322 ICBC PERSONAL BANKING

336 FIFTH THIRD BANK NORTHERN OHIO HEADQUARTERS

342 BANK IN PONSACCO

348 GUARANTY TRUST BANK

354 STANDARD BANK LONDON

366 HSBC REGIONAL EXECUTIVE FLOOR

372 KOREAN EXCHANGE BANK

378 INDEX

_Description

The client's guidelines for the project stressed on transparency of space, high flexibility in arrangement, modern, elegant look and direct contact with client.

Model outlet interior has been divided into three access areas significantly facilitating orientation of customers in the bank. Without using the traditional partition walls the designers managed to distinguish self banking, customer service area and background facilities. Ceilings on the whole area are covered by custom designed lamps tightly adjoining to each other. Cylindrical shape of the lampshades and blackness of the ceiling that increases impression of depth, both increase the feeling of space of the interior. Some lamps are used as ad carriers. Images placed at the bottom of the lampshade use the lamp lightsource to look similarly to citylight.

The package of branding elements designed by medusa also includes internet post, fun-place for kids, leaflet stands, a form of retractable lightbox with highlighted ad and multimedia pylon. For better customer orientation designers have created signage system in each room located on the floor.

ING BANK

Design Company_Medusaindustry

basic info   Designer_Wojciech Eksner, Łukasz, Zagała, Przemo Łukasik   Location_Poland   Photographer_Agnieszka Wawro, Milosz Jaksik

Co-authors Tomek Majewski, Daria Cieslak, Grzegorz Pietraszuk, Michał Sokołowski, Konrad Basan, Joanna Sowula

客户对此项目的要求是：设计要强调空间的透明度，布局安排要高度灵活性，要有现代感及优雅的外观，可与客户直接接触。

这个模型式的室内布置被分为三个基本的区域，并以客户为方向定义三个区域的功能，设计没有采用传统的分隔墙，但是仍然能分出自助区、客户服务区和背景设施等不同空间。整个区域内的天花板上满是专门设计的紧紧相邻的灯具。圆柱形的灯罩和黑色的天花板给人一种很有深度的感觉，同时增加了室内的空间感。有些灯具上还印着广告，更好地推广了银行。位于灯罩底部的图像利用光源到达这样一种效果：使其看似城市的灯光。

由 medusa 设计的一系列的品牌元素同样包括互联网、儿童玩耍区、放置传单的架子以及带有突出的广告和多媒体塔的可伸缩的灯盒。为了更好地为客户服务，设计师们已经在地板上的每个房间里都设置了引导标示。

OPTYMALNE POWIERZCHNIE	
STREF A SELFBANKING	15.0 m²
SALA OPERACYJNA	50.0 m²
POM. URZ. TRANSAKCYJNYCH	15.0 m²
BACK OFFICE	6.0 m²
POMIESZCZENIE TECHNICZNE	6.0 m²
POMIESZCZENIE SOCJALNE	10.0 m²
PRZEDSIONEK	3.5 m²
TOALETA	1.3 m²
POMIESZCZENIE GOSPODARCZE	1.3 m²
POKÓJ SPOTKAŃ	10.0 m²
RAZEM	118.1 m²

PLAN - MODERN VERSION

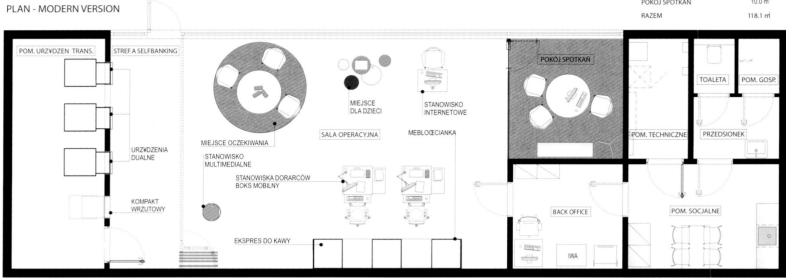

| SUFIT PODWIESZANY Z PŁYT G-K | ŚCIANA matowa farba lateksowa StoColor Latex 3000 kolor RAL 9003 | ŚCIANA NAD BLATEM ANEKSU mozaika ścienna Paradyż Albir Bianco 30x30cm | ŚCIANA W TOALECIE mozaika ścienna Paradyż Albir Bianco 30x30cm | ŚCIANA W POKOJU SPOTKAŃ matowa farba lateksowa StoColor Latex 3000 kolor pomarańczowy RAL 2004 | APLIKACJA GRAFICZNA LEW błyszcząca farba lateksowa StoColor Latex 9000 kolor pomarańczowy RAL 2004 | SUFIT KASETONOWY ARMSTRONG Plain 60x60 cm KOLOR BIAŁY |

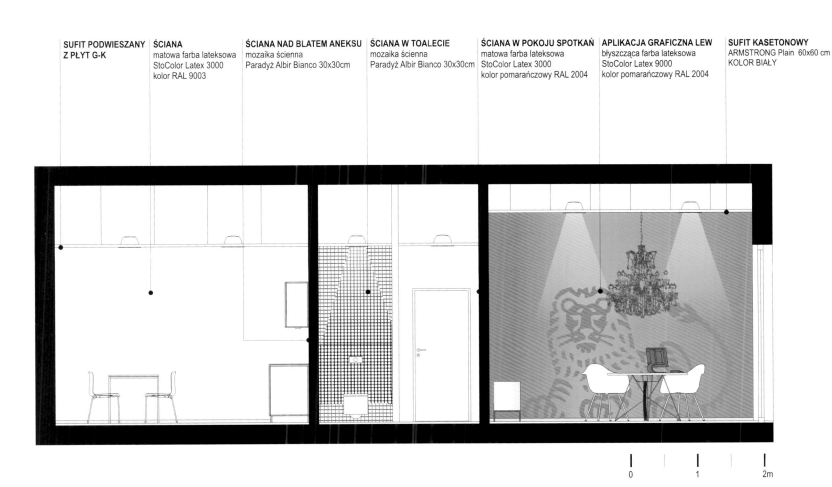

PLAN - TRADITIONAL VERSION

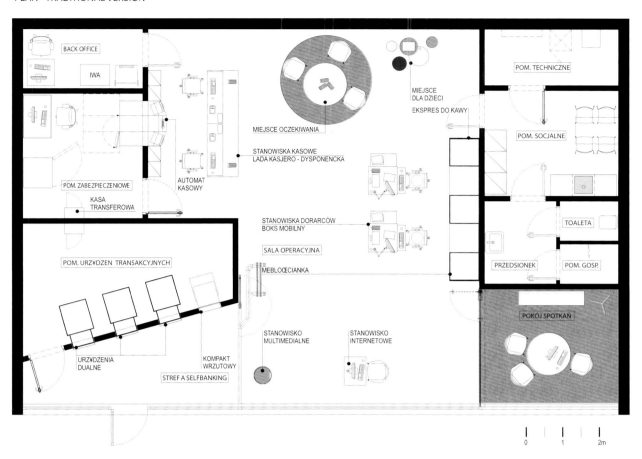

OPTYMALNE POWIERZCHNIE	
STREFA SELFBANKING	14.0 m²
SALA OPERACYJNA	85.0 m²
POM. URZ. TRANSAKCYJNYCH	16.0 m²
POM. ZABEZPIECZENIOWE	11.0 m²
BACK OFFICE	6.0 m²
POMIESZCZENIE TECHNICZNE	6.0 m²
POMIESZCZENIE SOCJALNE	10.0 m²
PRZEDSIONEK	3.5 m²
TOALETA	1.3 m²
POMIESZCZENIE GOSPODARCZE	1.3 m²
POKÓJ SPOTKAŃ	10.0 m²
RAZEM	164.1 m²

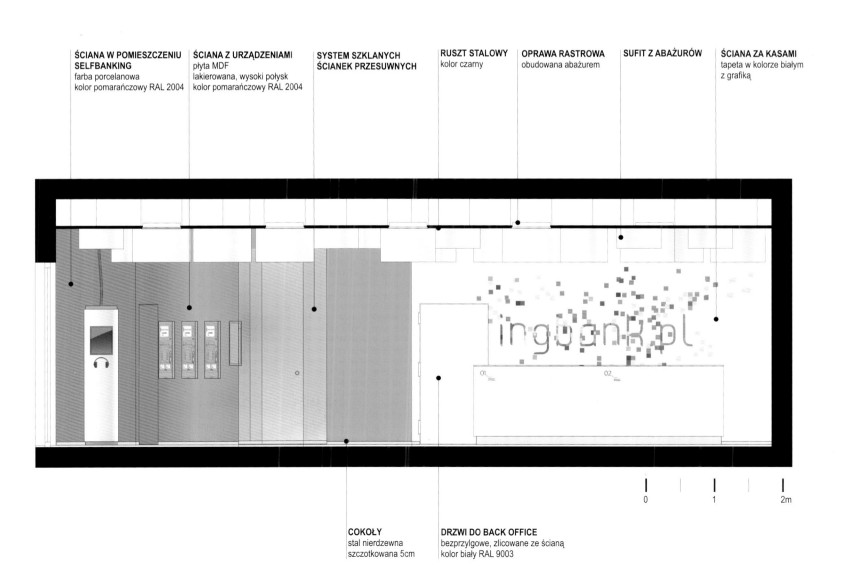

ŚCIANA W POMIESZCZENIU SELFBANKING
farba porcelanowa
kolor pomarańczowy RAL 2004

ŚCIANA Z URZĄDZENIAMI
płyta MDF
lakierowana, wysoki połysk
kolor pomarańczowy RAL 2004

SYSTEM SZKLANYCH ŚCIANEK PRZESUWNYCH

RUSZT STALOWY
kolor czarny

OPRAWA RASTROWA
obudowana abażurem

SUFIT Z ABAŻURÓW

ŚCIANA ZA KASAMI
tapeta w kolorze białym z grafiką

COKOŁY
stal nierdzewna
szczotkowana 5cm

DRZWI DO BACK OFFICE
bezprzylgowe, zlicowane ze ścianą
kolor biały RAL 9003

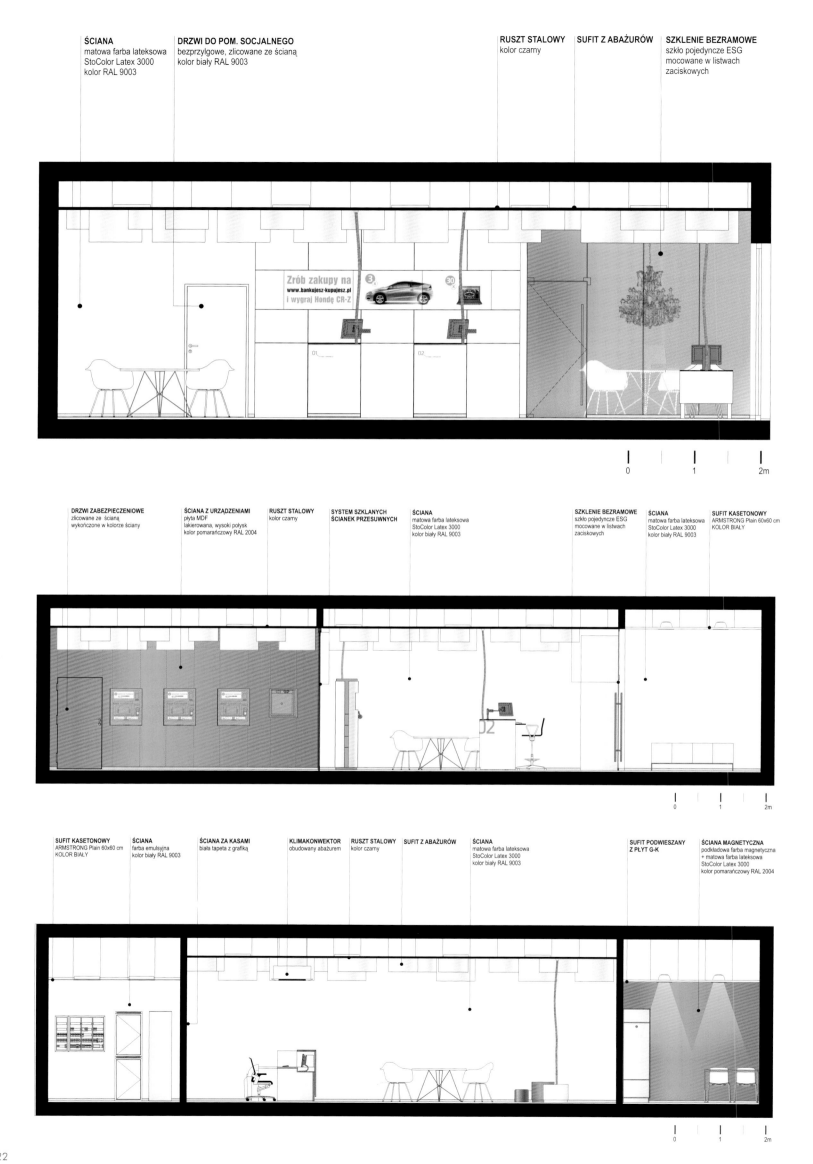

CORPORATE DEPARTMENT OF ING BANK ŚLASKI

Design Company_Robert Majkut Design

| basic info | Designer_Robert Majkut | Location_ul. Malczewskiego 45, Warsaw, Poland | Area_1,404m² |

_Description

The basis of the concept was the idea of creating new, abstract space for corporate customers sector, which is extremely important for the bank.

Solution
Taking into consideration functionality, solutions used concentrated on maximum effectiveness and perfect matching to the character and duration of meetings, required technological equipment and the way of their organization. Central, spacious, representative reception is also adjusted for the waiting room with its business TV channels, coffee corner for people who are waiting and also for those who are in the cabinets, and the communication network which enables easy access to all cabinets.

The client zone is linked directly with the back office. In the back office there are also cabinets of the management board and directors of particular departments and regions. Apart from the furnishing, the standard of the back office does not differ from the customer service zone.

This consequence of improving the furnishing standard is characteristic for the project. It should emphasize not only the characteristic way of brand image building but also care for the work quality, deeply rooted in the interior and culture of the company. In the back office there are comfortable armchairs for resting during informal meetings as well as equipment for individual or group work.

Formally, the project is based on the concept of free flow of waiving walls. Made of two layers, orange and white, they create a kind of stiff curtain freely covering or showing the entrances, passages, and doors in the whole customer zone. Free waiving and sloping wall gives the futuristic effect and diversify the design and intervention effect of the building. It creates the impression of sewing the "internal clothing" on the fixed building construction. Every hole in the structure shows its thickness, multilayer, additionally underlined by internal lightings. The same rule creates the internal island.

Colours which dominate the interior are the corporate ING colours. The usage of much higher intensity than in other branches, impose orange as the leading colour, supplemented with big amount of white and few hues of grey. This set of colours, enriched by lighting blue, visible mainly in the carpeting and furniture in the social infrastructure, breaks the warm energetic climate, balancing and making it calmer. The colour theme is seen in the whole space. White furniture, orange seats and walls, blue accents in the upholstering and carpeting create the perfectly harmonized wholeness.

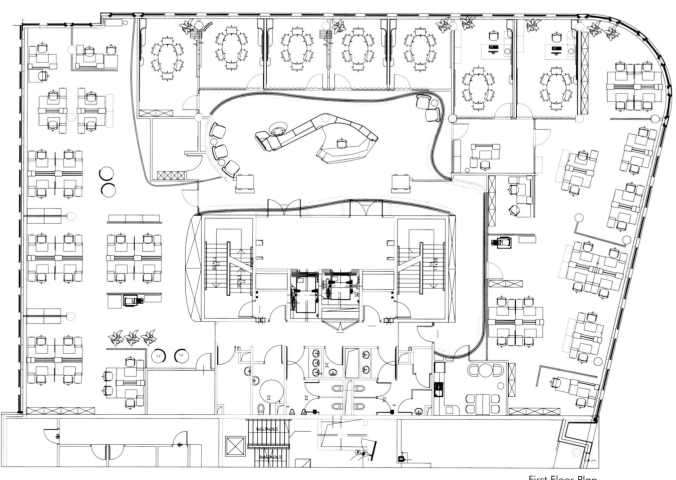

First Floor Plan

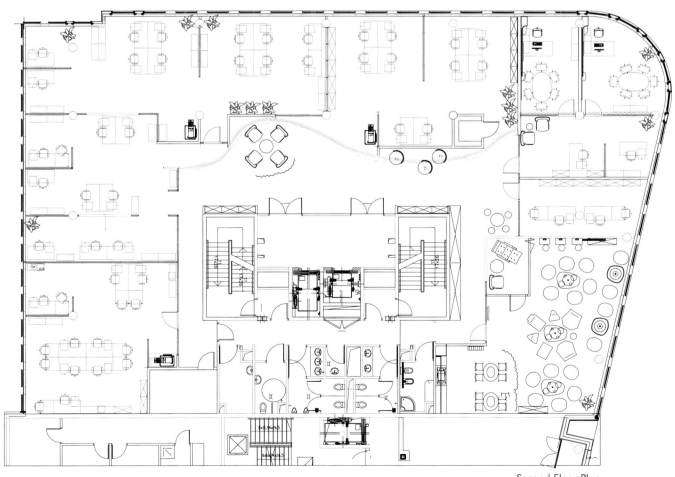

Second Floor Plan

这个概念的基础就是要为公司客户部门创造一个新颖、抽象的空间,这一点对银行来说是相当重要的。

解决办法

考虑到实用性,设计师所采用的方法主要集中在最大地提高有效性,使其与会议的性质及会议时间、所要求的技术设备和公司形式完全一致、完美相称。位于中心位置的宽敞的且具有代表性的前台也做了一定调整,改成了等候室,并配备了商业电视频道和咖啡角,以便其更好地为等待的客户和那些在小窗口的人们服务。同时调整了通讯网络,方便人们进入所有的窗口。

客户区与后勤部是相通的。后勤部里也有管理委员会办公室及某些部门和区域主管办公室。除了室内设备外,事务部门的标准和客户服务区也保持一致。

对于这个项目来说,改善设备标准的影响是典型的。它不仅要重视树立品牌形象大楼的独特方式,还要关心深深植根于公司文化和内部的工作质量。后勤部里有为非开正式会议的人们休息所准备的舒适的扶手椅,还有供个人和集体使用的设备。

正式地说,这个项目是基于这样一个概念:舞动墙的自由流动。它有两层,一层是橙色的,一层是白色的。它们创造了一种硬窗帘,可以随意地遮住或展示整个客户区的入口、通道和大门。随意摆动的舞动墙和倾斜墙给人一种到了未来的感觉,还丰富了设计,增强了建筑的美感。它让人有种看到了固定建筑的"里层衣服"的感觉。从结构中的每个孔洞里,我们都可以看出它非常厚,且是多层的,然而在内部灯光的映衬下,显得额外突出。同样的理论也应用于内部设计。

内部的主要颜色是荷兰国际集团的公司颜色。因为与其它分公司相比,这次的项目使用了多种强亮度的颜色,所以,只能将橙色作为主色调,再配上大量的白色和少量的灰色色调。这一整套颜色在蓝色的照明灯下,映衬地更为亮丽,您能在作为基础设施的地毯和家具上看到这些,它打破了温暖、活泼的氛围,使气氛更为平和、让整个环境更加平静。这样的颜色主题充斥了整个空间。室内装潢内白色的设备、橙色的座椅和墙面、大面积的蓝色和地毯创造了一个相当和谐的环境。

31_

CHEBANCA!

Design Company_**Crea International**

basic info　　Designers_Marco Michele Rossi, Andrea Borsetto　　Location_Italy

Crea design concept for CheBanca is "natural tech". The layout of CheBanca is organized to remember the logical organization of the solar system with the client ideally at the centre of it. The natural tech of CheBanca means ethic and transparency of a world that does not deceive. The yellow color that permeates the environment reminds of the sunshine light, the aniline treated wood suggests a straightforward approach, and the material printed with the honeycomb texture casts a friendly atmosphere.

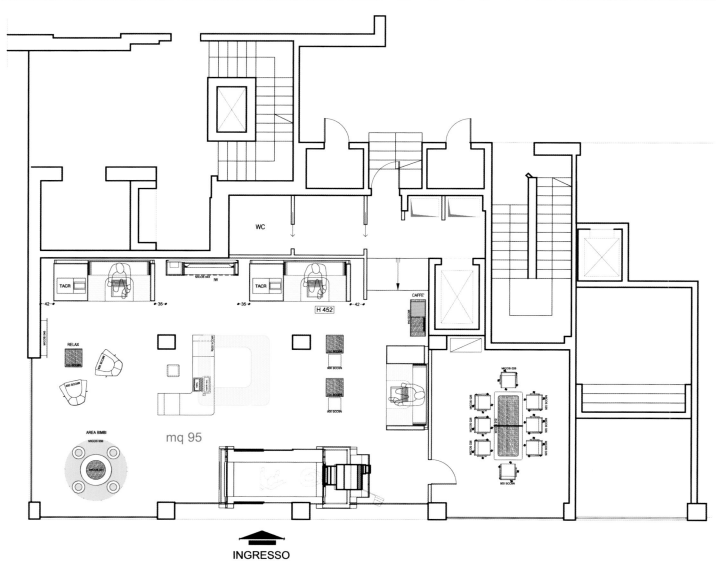

INGRESSO

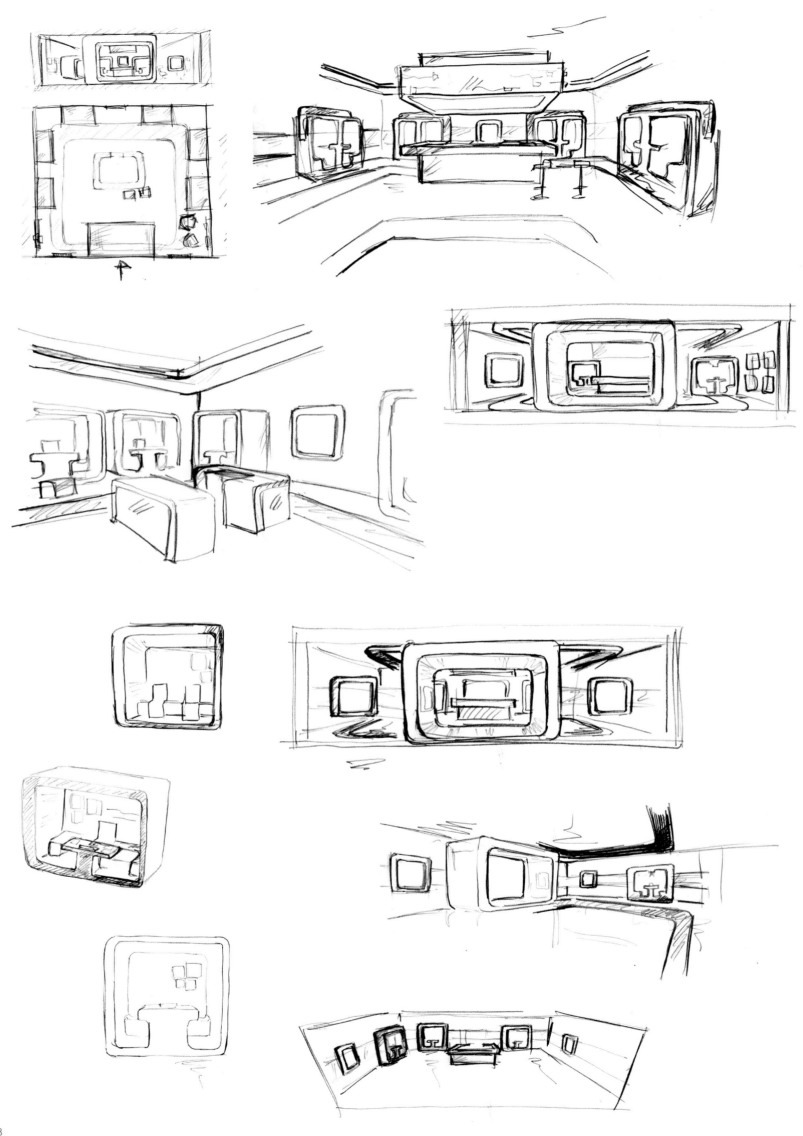

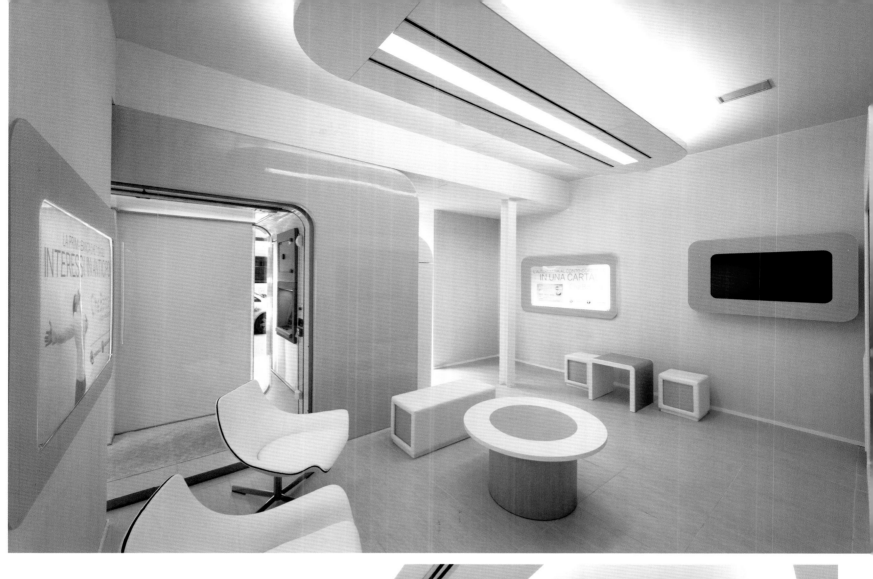

Crea国际团队给CheBanca银行的设计理念是"自然的高科技"。CheBanca银行的布局参考太阳系的逻辑排列强调顾客始终是核心的理念。它的自然高科技意味着真实世界的职业道德和透明度。漫延于室内的黄色让人有一种被阳光包围的感觉,用苯胺处理过的木材暗示了一种直截了当的方法,印有蜂窝纹理的材料则使您置身于友好的氛围中。

NATIONAL BANK OF OMAN

Design Company_**Allen International**

basic info | Designer_**James White** | Location_**Oman**

_Description

Allen international was selected to develop a new brand identity and retail experience for NBO, the second largest bank in Oman with over 50 branches. The major challenge was to develop a strategy that clearly positioned the bank and to assist NBO in understanding their local roots but with regional and international connections.
Through the development of a brand and retail strategy allen international quickly focused on some key strategic thoughts that positioned NBO as an "Omani bank for Omani" and a Brand DNA® of being an "Enterprise for a new generation". The brand evolved as the spirit of Oman through colors and soft flowing forms evocative of wind and incense. The retail experience opens from the welcoming nature of Omanis and the "majalis" – a place where people meet and converse. A relaxed central space has been created with integrated merchandise surrounded by areas of privacy and consultation.
Pilot branches are currently being implemented and tested.

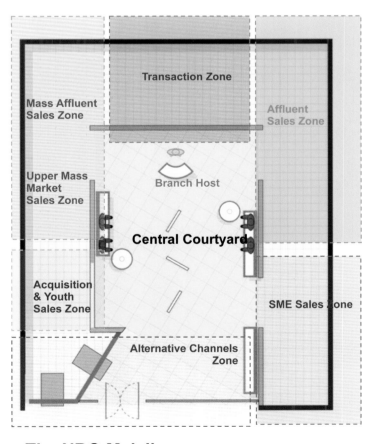

The NBO Majalis

艾伦国际设计公司应邀为阿曼国际银行开发一种全新的品牌形象和零售经验，该银行是阿曼第二大银行，拥有50多家分行。设计面临的主要问题是建立一套战略方案来定位该银行，并以地区和国际关系来帮助阿曼银行认识其当地根基。

随着其中一种品牌和零售战略的发展，艾伦国际很快把重点放在一些核心战略思想上，那就是把阿曼国际银行定位为"阿曼人自家的银行"以及"新生代企业"这样一种品牌基因。这种品牌通过色彩和柔软的流动形式，完美演绎了阿曼人的精神，使人们陶醉于微风和香气之中。它的零售经验则从两个方面展开：阿曼人热情好客的品性和"majalis"——人们见面和交谈的地方。银行还创造了一个休闲的中央空间，并摆着被私人区域和咨询区围绕的综合性商品。

试点分行现正处于运行和检测阶段。

BARWA BANK

Design Company_ **Crea International**

basic info | Design director_ Libero Rutilo | Concept supervisor_ Massimo Fabbro | Architect_ Nicola Golfari

_Description

The Barwa Bank branch design concept has been one of the most challenging projects that Crea International has ever developed: designing the most progressive Islamic bank of the future, showcasing modernity and coolness. The objective was to propose an environment aimed at making customers more and more familiar with the most innovative technologies, a bank thought where people would feel comfortable and welcome.

Through the methodology of Physical Brand Design, Crea International team approached the project in a very logical and structured way: "In the first place, we look into the history and traditions of the country to get familiar with the components Qatari people felt very belonging to their culture and were proud of. We also analyzed throughout the mission of Barwa Bank and the values it stands for, and finally we looked at the banking models both in Western and Middle East countries to build a strong point of difference and a gap with the current models, as claims Viviana Rigolli, strategy director of the project."

To ensure real distinctiveness versus such current banking models and to build the most innovative bank, Crea International designed a new service standard: only a central banking area hosting multifunctional comfortable workstations where the bank assistant can seat close or in front of his client looking together at touch screen table, where all banking functions can be performed with total transparency and almost paperless. The final result is a delicate balance between simplicity, intuitive space fruition of the service model and the warm environment.

Strategy director_**Viviana Rigolli** Lead graphic designer_**Sonia Micheli** Graphic designer_**Giuseppe Liuzzo** Location_**Qatar**

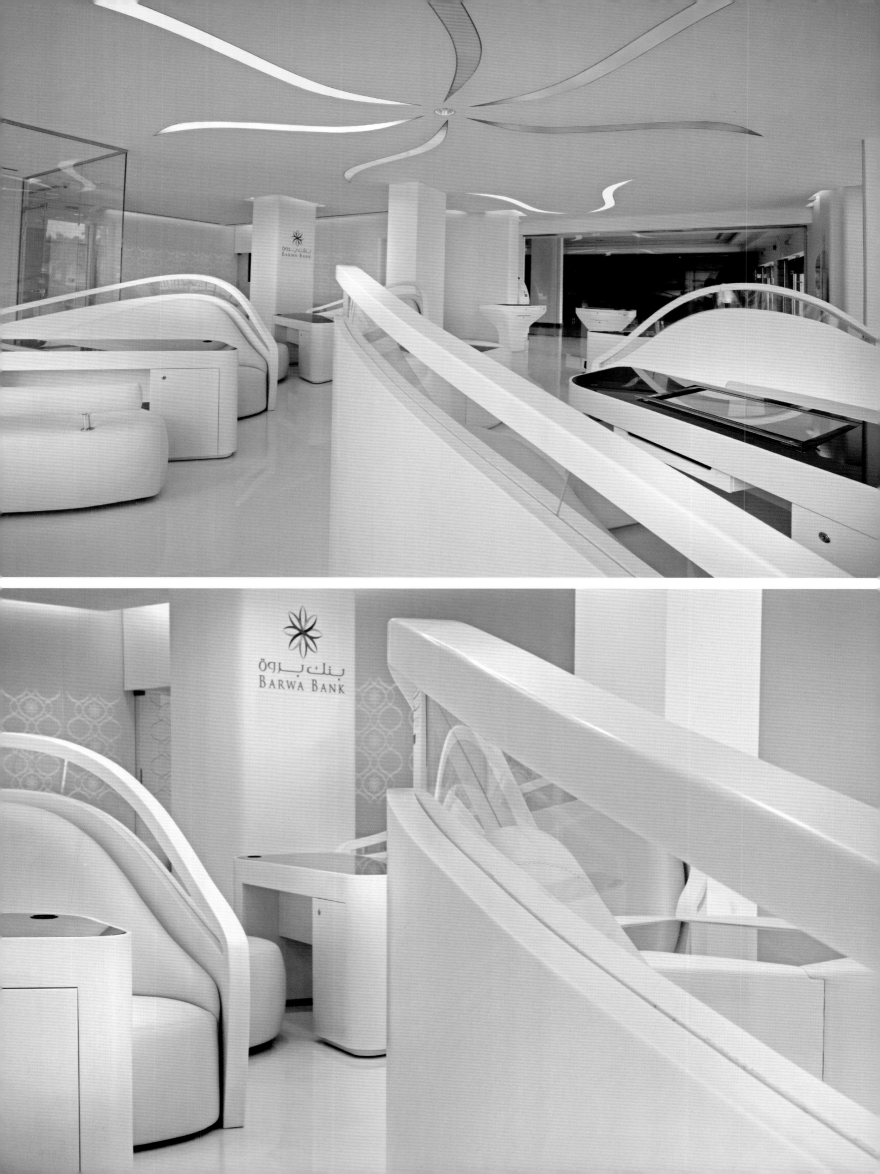

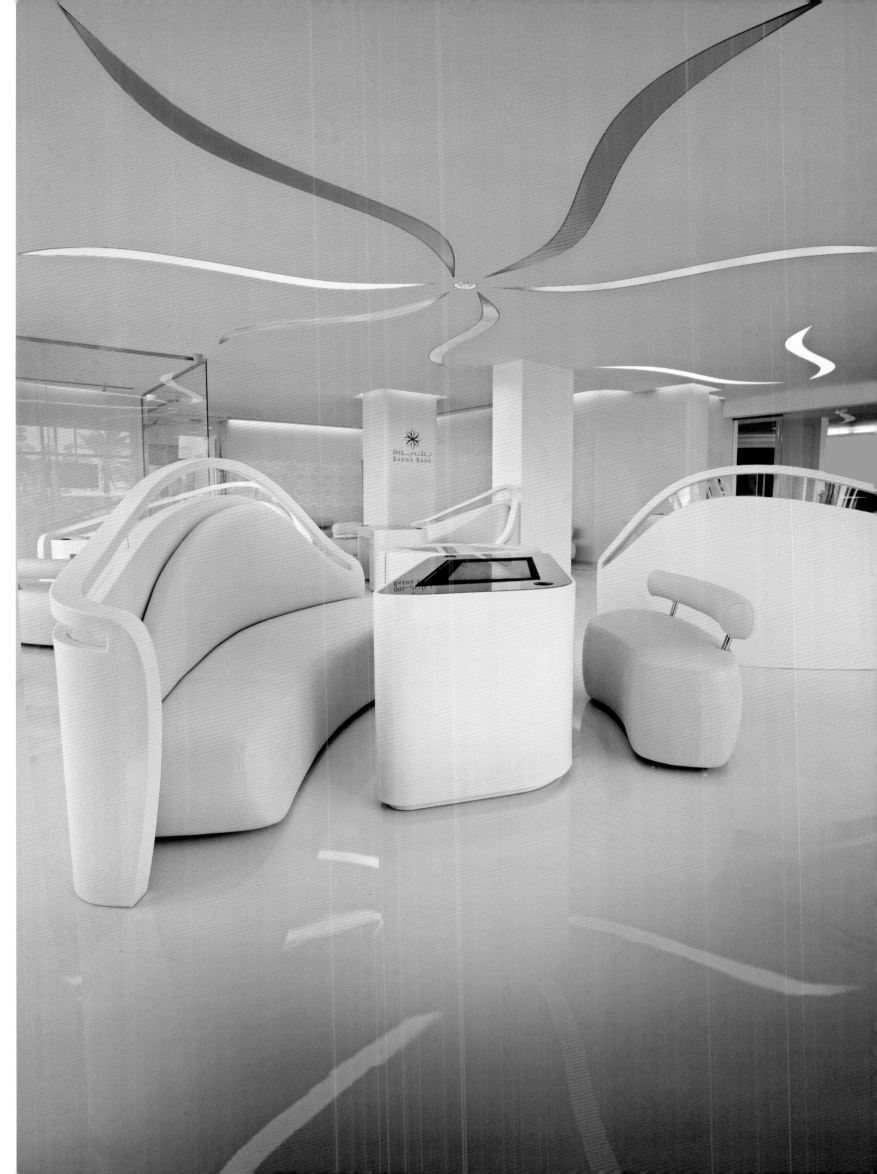

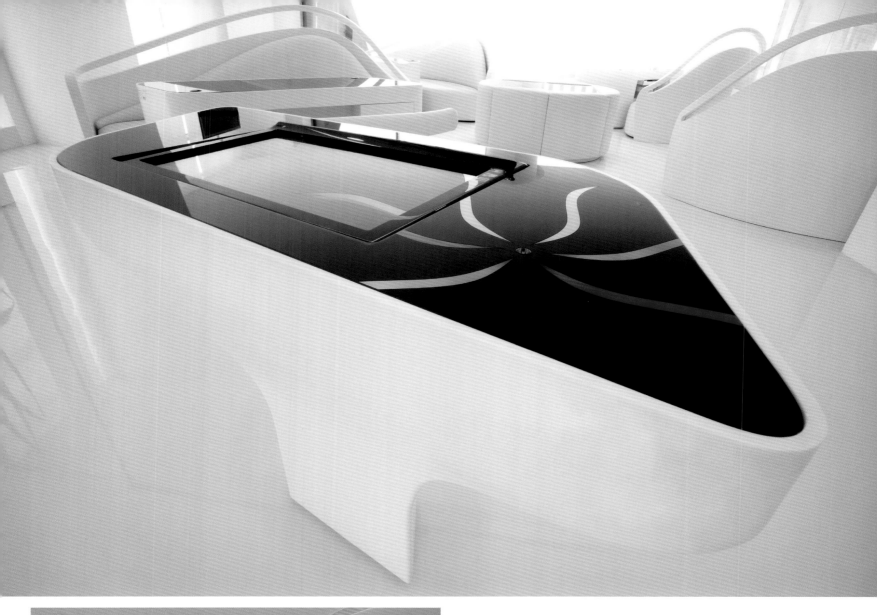

Barwa银行分行的建筑设计理念是Crea国际设计公司有史以来所承接的最具挑战性的建筑项目之一：设计一个最具创新精神的伊斯兰银行，并展现出其现代感及冷静、沉着的精神。设计的目的是打造一个以客户为主的室内环境，让客户能够越来越熟悉最创新的科技；同时打造出一种让客户宾至如归的银行理念。

通过体育品牌设计的方法论，Crea国际团队用结构条理分明的方法来设计这个银行大厦。该银行建筑项目的策划总监Viviana Rigolli认为："首先，我们要了解这个国家的历史和传统，逐渐熟悉那些能让卡塔尔人觉得有归属感和自豪感的建筑元素。同时，我们还分析了Barwa银行的服务宗旨和价值观，然后浏览了西方及中东一些国家银行建筑的先例，以便能够打造出与现有建筑不同的设计。"

为了确保与现代的银行建筑不同，且是一个创新型银行，Crea国际为该分行制定了一个新的服务标准：只规划设计一个中央服务区域集多功能、舒适的工作岗位，银行协理员的座位可以接近客户或是在客户前面，二人能同时看到可触屏桌，这样所有的银行业务都能够透明化、无纸化操作。服务区域空间效果简捷、直观，办公环境优雅温馨，二者间达到了微妙的平衡。

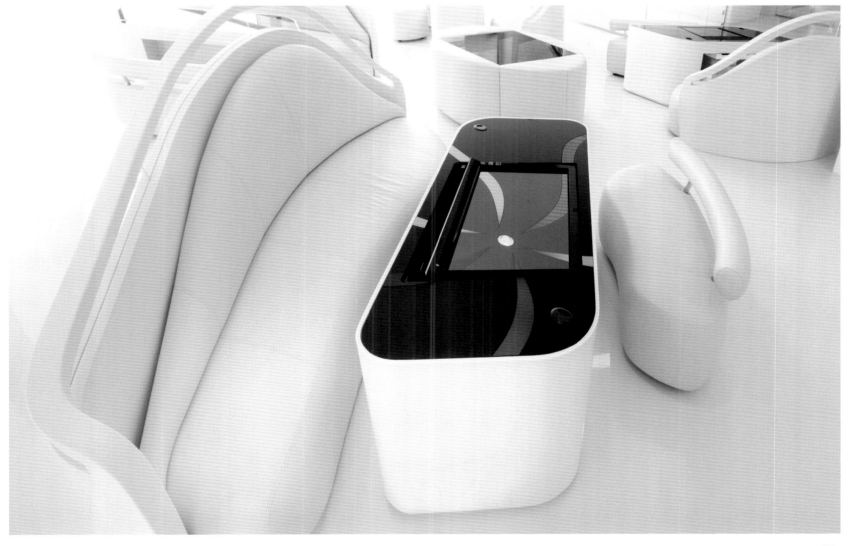

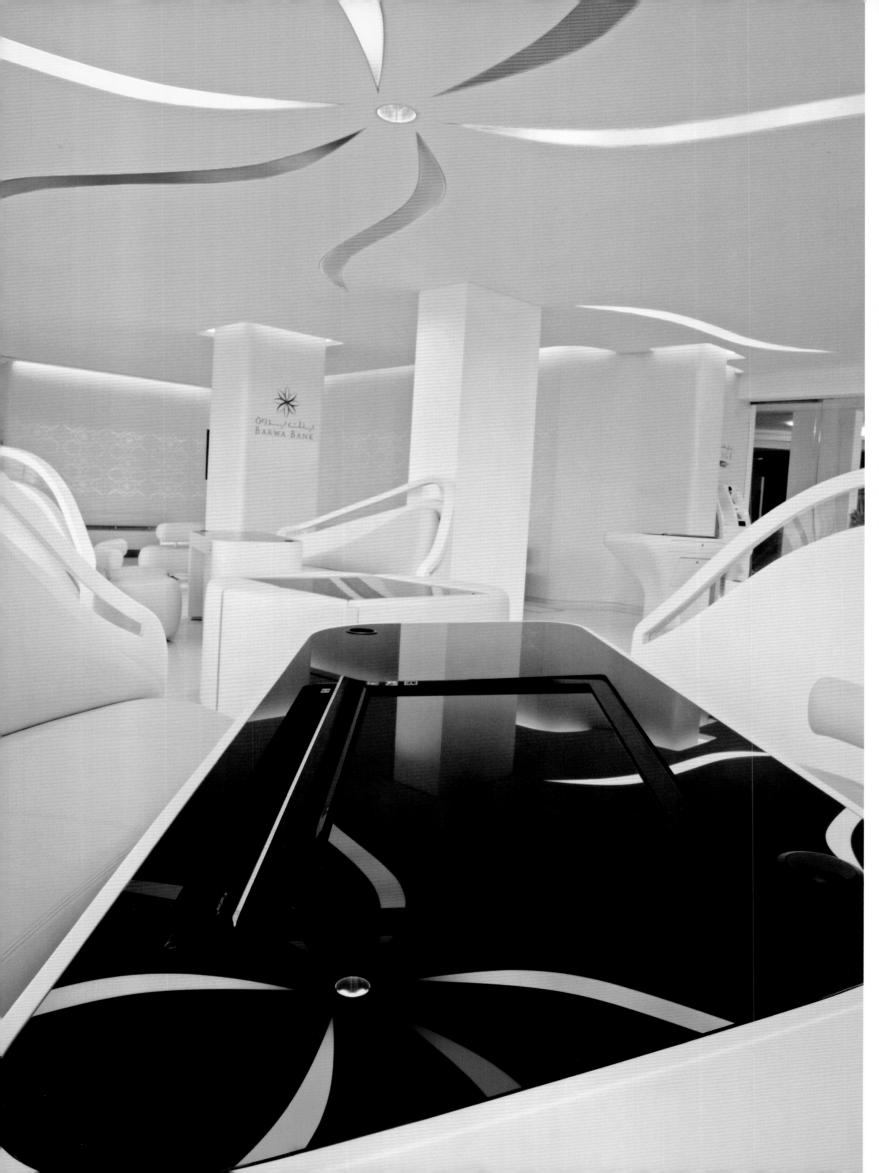

_Description

The ground and first floor of Deutsche Bank have a unique elliptical cut-out, right in the centre, connecting the two spaces. And the facade on these floors has glazed paneling facing the road, making this an obvious choice for the branch and public interface operations. The business unit on the first floor is internally connected by a staircase designed specifically for the purpose. Innovatively designed in steel and white agglomerate marble, the spine and fin structure creates a striking design element. The bank's corporate colors of blue and white are cleverly combined throughout the interiors. Visitors can see a blue square, inspired by the Deutsche Bank logo. The blue square is creatively suspended from the ceiling of the first floor.

One of the main areas of importance in every bank is the IT centre. In the Deutsche bank, the Server and Equipment Room (SER) housing the network server and the communications hub are placed on the second floor. Being the heart of the banking operations, the central location helps cuts down IT costs to a great extent and also increases efficiency. The challenge was to connect various floors without damaging the existing structure; also as per the bank's IT standards, the connections had to be seamless. To achieve this, a unique ducting system was designed. This was made from powder coated M.S. box sections, positioned externally, entering on each floor and further routed via aluminum floor raceways. The ducts were designed to be moisture proof.

Exploration of the second and the third floors reveal zones for other corporate bank functions. Here the Architects experimented with the finishes, combining the corporate colors with a touch of rich veneers, matt steel and carpets to achieve a look that was dramatic and adhered to the global design principles prescribed by the bank. The designing and coordinating of various services such as the electrical, HVAC and security systems was achieved after carefully studying each layout, superimposing them on each other and making necessary modifications without disturbing the basic functioning. Today the seamless functioning of the Deutsche Bank in Pune, is proof of the excellent architecting solutions achieved through meticulous planning and execution.

DEUTSCHE BANK

Design Company_**JTCPL Designs**

Area_1,394m² Location_Maharashtra, India

basic info

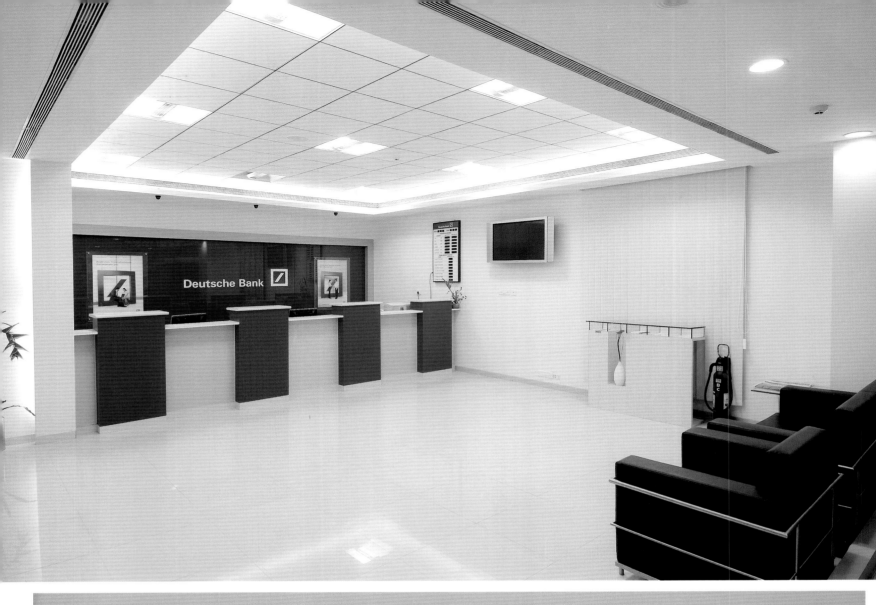

　　德意志银行大厅的正中央坐落着一个独一无二的镂空椭圆形球体,连接着银行的底楼和二楼。每一层的立面上都铺满了玻璃嵌板,正对着道路,这种效果正是该分行和公共运行界面的不二选择。二楼的业务部门由一个特殊设计的楼梯专门连接到室内。钢铁和白色大理石的创新性设计使得骨结构和鱼翅结构成为引人注目的设计元素。蓝色和白色的企业色彩被巧妙地结合起来,贯穿到整个内部结构中。来访者可以看到一个蓝色的正方形别出心裁地悬挂在二楼的天花板上,它是来源于德意志银行的标志。

　　计算机技术中心在每一家银行都扮演着极为重要的角色。在德意志银行中,服务器和设备机房控制着网络服务,而通信中心则设置在三楼。中央位置作为银行金融业务的核心,它不仅在很大程度上减少了信息技术的成本,更提升了效率。计算机技术中心所面临的挑战就是在保持原有结构的基础上对每一层通讯都形成有效的连接。对于银行的信息技术的标准来说,这种连接必须是天衣无缝的。为了达到这一目标,需要设计一种独特的管道系统。它是由黑色亚光处理的最大应力方形管制成,并设置在银行外部,通过铝制地管道输送进入每一层楼。此外,管道应是防水的。

　　大楼的三楼和四楼将被设计出来提供给其他股份银行使用。因此,建筑师们使用表面材料,如用一些厚单板、马特钢和地毯等进行试验,与企业色彩结合,营造一种与银行规定的全球设计原则相吻合的外观。各种服务的设计和协调包括电力、高压交流电以及安全系统在不打乱基本功能的情况下,通过对每个设计的认真研究、相互叠加以及略加修改后得以完成。如今,在设计师们一丝不苟的策划和实施下,位于普纳的德意志银行的运行情况无懈可击,被证明为完美的建筑方案。

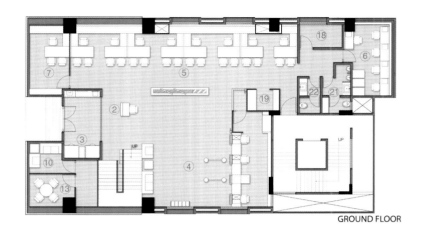

GROUND FLOOR

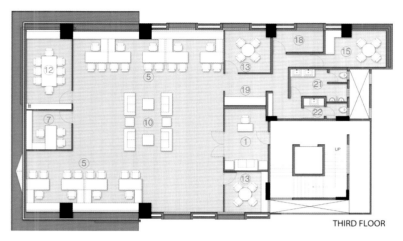

THIRD FLOOR

FIRST FLOOR

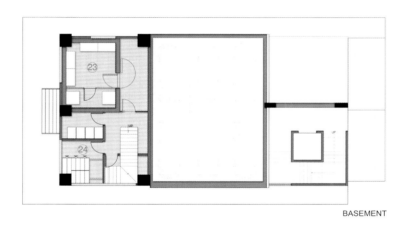

BASEMENT

SECOND FLOOR

1	RECEPTION AREA		13	INTERNAL MEETING ROOM
2	METER GREETER		14	CAFETERIA
3	ATM		15	PANTRY
4	TELLER		16	SERVER ROOM
5	FRONT OFFICE		17	UPS ELECTRICAL
6	BACK OFFICE		18	STORAGE
7	AREA MANAGER		19	UTILITY
8	BMS ROOM		20	PRINT STATION
9	IES ROOM		21	GENTS TOILET
10	EXECUTIVE LOUNGE		22	LADIES TOILET
11	LOUNGE / BREAKOUT		23	VAULT ROOM
12	CONFERENCE ROOM		24	COMPACTOR STORAGE

HSBC SHANGHAI

Design Company_**Robarts Interiors and Architecture**

| basic info | Location_Shanghai, China | Area_41,806m² | Photographer_Owen Raggett |

_Description

In 2010, HSBC moved into their new China headquarters in Shanghai. The 41,806-square-meter offices are located in HSBC Tower, right next to the Shanghai World Financial Center, with spectacular views across the river to Shanghai's iconic Bund, which was home to the original HSBC bank building, recalling the bank's founding in1865.
Initially occupying 16 full floors, the office is home to over 2,000 employees, a substantial IT datacenter and a sizable trading floor. Whether in the impressive client areas or in the working office that client basically don't see, the whole project is designed to reinforce the HSBC brand, making use of quality materials and furniture that will stand the test of time.
The design employs an understated color palette, defined by the HSBC brand's predominantly black, white and grey tones accented with careful use of HSBC red and polished stainless steel details (such as the 7.6 meters high contemporary screens evoking the bank's logo style). Water features also provide white noise to key client areas. Deliberately less machined and neutral than the interiors of the bank's well-known headquarter projects in Hong Kong and London, the Shanghai project interiors bravely use more wood and felt upholstery to keep the environment warm and welcoming whilst staying professional and clearly modern.

To reduce the carbon footprint of the project, all wood is bamboo from southern China, including mulched bamboo veneers on furniture customized for the project by Haworth. Beautiful felt upholstered wall panels also remind one of Inner Mongolia in northern China.
The project is receiving LEED CI Gold certification from the US Green Building Council, so materials, lighting, and mechanical systems have all been through rigorous review to comply with the world's best-in-class standards for environmental sustainability. The professional team that designed and delivered this project worked for two and a half years, since the building was still not out of the ground when they commenced their work.
The delivery of the project involved a multi-faceted approach that engaged all key constituents of HSBC – Senior Management, Business Unit Champions and a cross section of the bank's staff to determine the goals and requirements for the project. The design team also studied HSBC's existing facilities and standards – including site surveys at both HSBC's landmark locations in Queen's Road and Canary Wharf. This process ensured that the project adhered to HSBC's global standards whilst responding to the unique nature of the future occupants in Shanghai.

  2010年汇丰银行入驻位于上海的新的中国总部大楼。这个面积为41806平方米的办公室位于汇丰大厦，右边紧挨着的是上海环球金融中心，在这里可以看到黄浦江边上海外滩的壮观景色，同时这里还是汇丰银行大楼的原址，这不禁让人想起了1865年银行刚刚建立的场景。

  起初办公楼有16层，是巨大的信息技术资料中心，也是相当大的贸易中心，可容纳2000多名员工。无论是在让人印象深刻的客户区还是在客户基本看不到的办公区，整个建筑项目的设计都是为了强化汇丰银行的品牌，充分利用那些经得起时间考验的优质材料和设备。

  该设计采用一种朴素的色调，这主要是因为汇丰银行的招牌的主要颜色是黑色、白色和灰色，此外设计师还非常注重汇丰银行红色细节和抛光不锈钢细节的使用（如与银行标志的风格相呼应的7.6米高的当代屏幕）。水文要素使得主要客户区免受噪音的影响。与著名的香港和伦敦总部的室内相比，上海总部显得不那么机械化和中性化，相反其室内大胆地使用木材和很有感觉的室内装潢品，从而在保持一个专业、清晰的环境的同时创造出一种温暖、吸引人的氛围。

  为了减少这个项目的碳排放量，所有木材都是从华南地区选购的竹子，包括由哈沃思为这个项目定制的用于设备上的覆膜竹胶合板。美丽的修饰墙板使人想起了中国北方的内蒙古的墙板。

  这个项目获得美国绿色建筑协会LEED金牌认证，因此为了达到世界级水准的可持续发展目标，所有的材料、照明设备和机械系统都已经经过严格的审核。专业团队在大楼还未出土时便接受了这个工作，设计到最终移交这个项目共花费了两年半的时间。

  这个项目的完成采用了各种方法，参与这个过程的包括汇丰银行所有主要部门——高层管理人员、业务部门和各部门员工，通过这种方法来决定项目的目标和要求。设计团队也研究了汇丰银行现有的设施和标准——包括对坐落于皇后大道的汇丰地标性建筑和位于金丝雀码头场地的研究和调查。这个过程确保这个项目在坚持汇丰国际标准的同时也能为上海未来居住者创建一个独一无二的自然环境。

OPEN LOUNGE
— RAIFFEISEN BANK

Design Company_ NAU Architecture, Drexler Guinand Jauslin Architekten

basic info | Area_ 400m² | Photographer_ Jan Bitter

_Description

Raiffeisen's flagship branch on Zurich's Kreuzplatz dissolves traditional barriers between customer and employee, creating a new type of "open bank". Advanced technologies make banking infrastructure largely invisible; employees access terminals concealed in furniture elements, while a robotic retrieval system grants 24 hour access to safety deposit boxes. This shifts the bank's role into becoming a light-filled, inviting environment — an open lounge where customers can learn about new products and services. This lounge feels more like a high-end retail environment than a traditional bank interior. Conversations can start spontaneously around a touch screen equipped info-table. The info-table not only displays figures from world markets in real-time, but can be used to interactively discover the history of Hottingen, or just check the latest sports scores.

Elegantly flowing walls blend the different areas of the bank into one smooth continuum, spanning from the customer reception at the front, to employee workstations oriented to the courtyard. The plan carefully controls views to create different grades of privacy and to maximize daylight throughout. The walls themselves act as a membrane mediating between the open public spaces and intimately scaled conference rooms. Portraits of the quarter's most prominent past residents like Böklin, Semper or Sypri grace the walls, their abstracted images milled into Hi-macs using advanced digital production techniques. While intricately decorative, the design grounds the bank in the area's cultural past, while looking clearly towards the future. The Info-table in the lounge also offers customers a chance to check the latest stock quotes, or the daily news.

The touch screen also allows one to learn about the history of the area. Most importantly it acts as a magnet for client and advisor to meet around casually, before deciding if they should move to the adjoining meeting rooms for further discussion. Displayed information is updated real-time from the web through programmed RSS feeds. New figures pop up like animated soap bubbles.

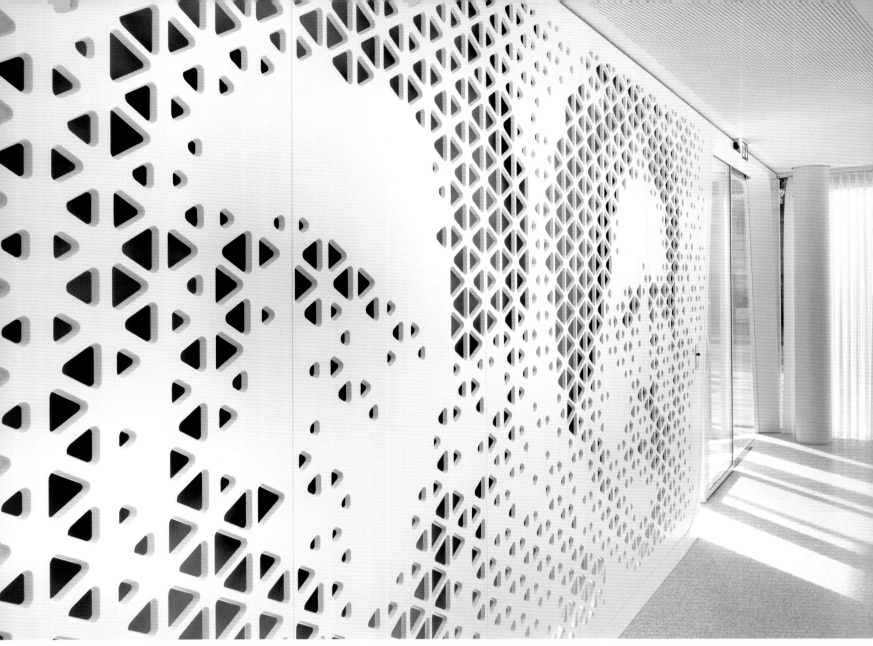

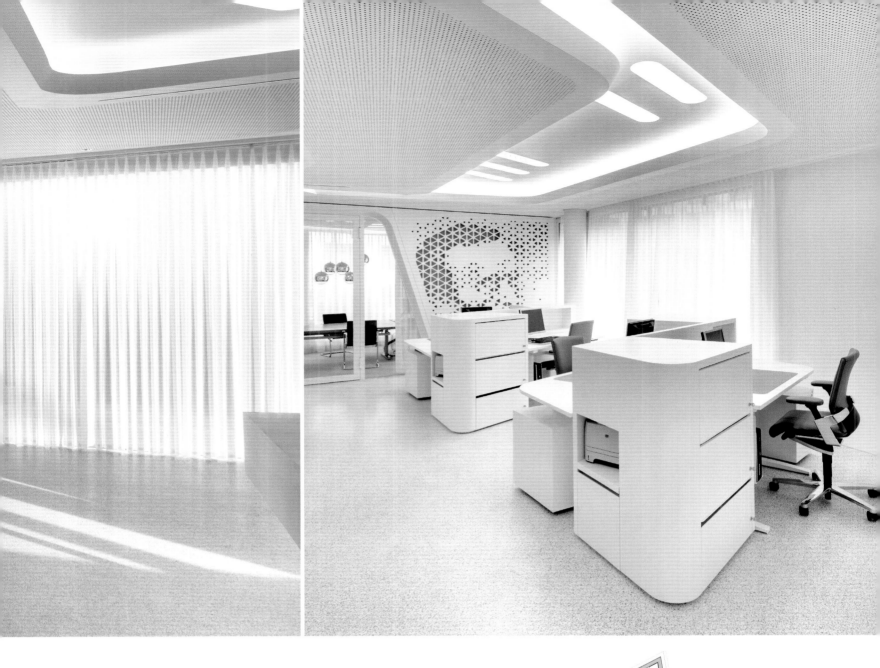

LEGEND

01 Entrance / ATM
02 Safety Deposit Access
03 Lobby / Reception
04 Cash Desk
05 Loug / Info-table
06 Meeting Room
07 Offices
08 Head Office
09 Break Room
10 Secure Zone

来富埃森银行在苏黎世Kreuzplatz的旗舰分行解决了客户和员工之间的传统障碍，创造了新型的"开放银行"。先进的技术使大量的银行基础设施变得无形；员工接入终端机被隐藏在家具中，取而代之的是机器的检索系统用来保证24小时接入保险箱。这把银行的角色转移到一个充满光亮、吸引人的环境中来，一个供客户了解新产品和服务的开放的休息室。这休息室比起传统的银行内室更像是一个高端的零售业环境。对话就可以自发围绕所装配的信息表触摸屏展开。信息表不仅能实时地展现世界市场的数据，而且可以用交互的方式发掘Hottingen的历史，或者是核对最新的体育比分。

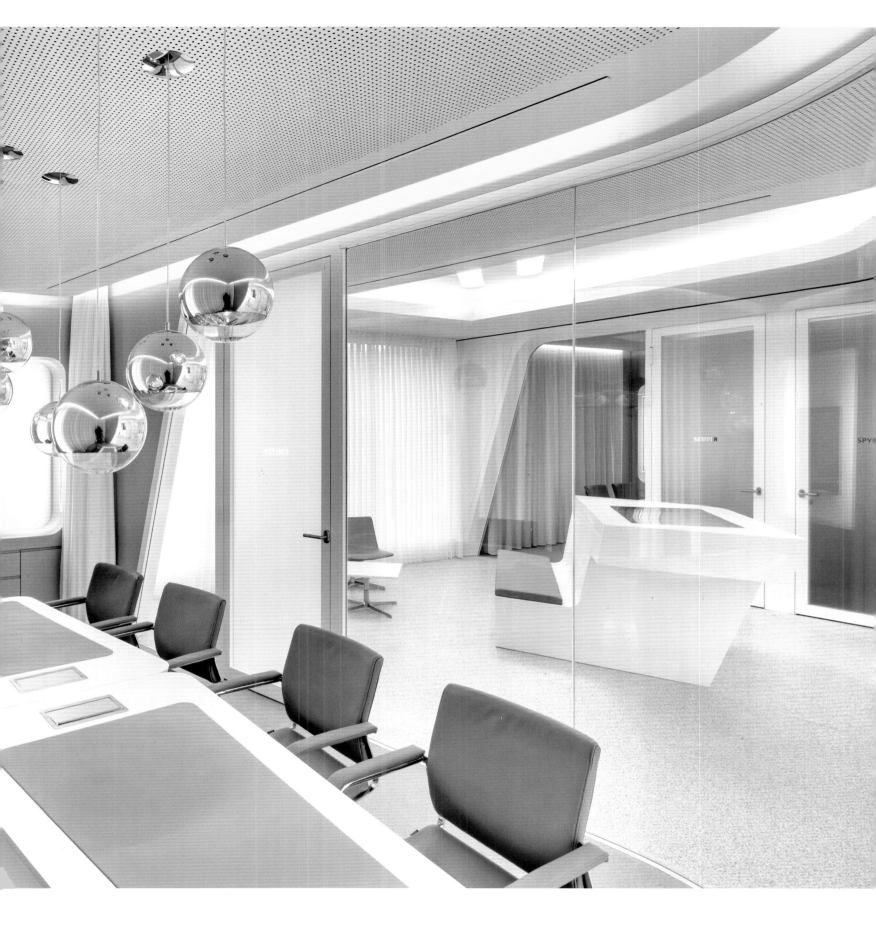

　　高雅流动的墙将银行的不同领域融合成一个连续体、涵盖了前方的客户接待处和面向庭院的员工工作站。该设计仔细控制视图去创建不同等级的私密性并最大限度地引进自然光。墙体本身则用作划分开放领域与私人空间会议室的中间层。本地区的最优秀的居民，如Böklin、Semper或Sypri，他们的肖像画通过高端数字化技术展现在数字电视上，使墙变得更加优雅。同时，错综复杂的装饰设计使银行在过去的文化基础上清晰地面向未来。休息室的信息表还使顾客有机会浏览最新的股票行情或每日新闻。

　　此触摸屏也能够让您了解该地区的历史。更重要的是，它像是磁场一样，在客户和顾问决定是否要移步毗邻的会议室作进一步讨论之前，可以吸引他们在此进行随性的会见。通过编程的RSS的网页实时更新显示的信息。更新的数据像动态肥皂泡一样自动弹出。

LLOYDS BANKING GROUP

Design Company_Allen International

Designer Nic Preece   Location United Kingdom

_Description

A thoroughbred amongst banks with over 11 million customers, Lloyds Banking Group is determined to retain its position as Britain's most loved financial institution. While others falter, Lloyds Banking Group remains nose and tail ahead of its competition.

Lead sponsors of the 2012 Olympics, a new national advertising campaign and a declared intention to partner its customers "through their life's journey". Lloyds Banking Group appointed retail design specialists, allen international, to change the face of their High Street branches. allen international developed a fresh contemporary look for the Lloyds Banking Group estate comprising of over 3,000 branches. The retail design strategy was based on a "Britishness" theme communicating a caring, creative and everyday Britain that was reflected in the new color palette, the materials and furnishings and finally the communications.

The then Head of Retail Banking for Lloyds Banking Group commented that the warehouse pilot branch presented by allen international, was the best conceptual branch she had ever seen.

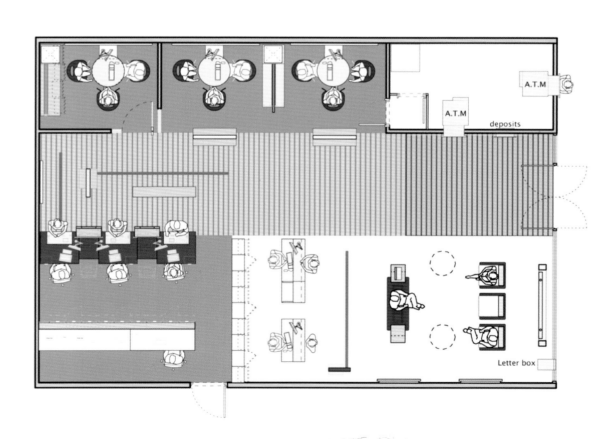

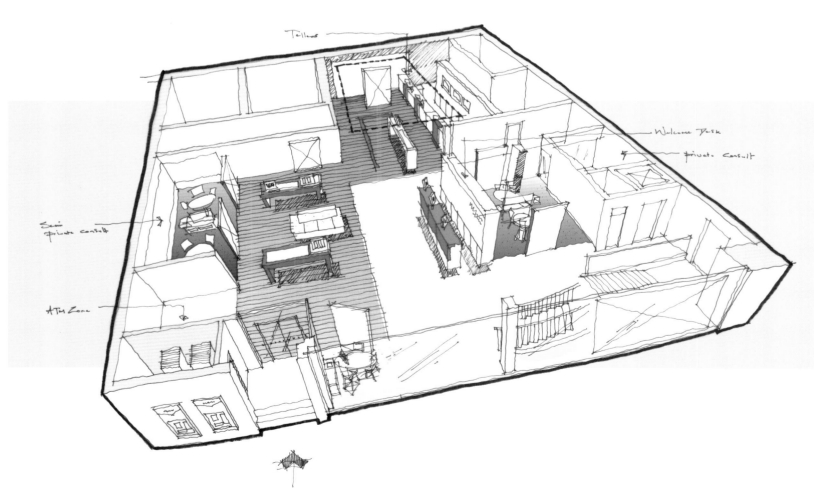

　　劳埃德银行集团有着1100万客户，是银行业中最为卓越的一家银行，它有决心保持其作为英国最受人喜爱的金融机构的地位。其他银行摇摇欲坠的时候，劳埃德银行集团面对竞争，仍然不动摇，继续努力。

　　一场新的全国性广告活动拉开帷幕，2012年奥运会的赞助商们以"伴您人生旅途"的标语向其顾客公开表明他们的立场。劳埃德银行集团委托艾伦国际——零售设计方面的专家来改变其位于繁华商业街上的分行的外观。艾伦国际为有3000多分行的劳埃德银行集团分行设计了一个新颖而具有当代气息的外观。

　　这个零售设计战略建立在"英国人风格"这一主题之上，这一主题体现了关爱人、具有创意的生活中的英国，而新的颜色搭配、材料、家具及最后的沟通和交流都集中反应了以上那些特点。

　　那时的劳埃德银行集团零售银行业务部门的主管认为这个由艾伦国际缔造的仓库式试验分行是她见过的最好的概念银行。

_Description

Standard Chartered Bank has a presence in 88 countries with retail branches in 39 of them across North and South East Asia, Africa, UAE and the Indian Subcontinent.

Standard Chartered approached allen international with the challenge of developing an universal branch standard that consistently delivered the bank's brand promise, with controllable costs, quality of materials and layout. Having done this allen international had to create a manual to support and underpin this.

Allen international created a strategy focused on a high level of human interaction supported by delivering the Attract, Build, Capture retail communications through the use of in-branch digital technologies. Communication to customers is delivered through managed digital media screens where customers can browse, print, e-mail product and service data through interactive terminals. Customers and staff can interact over long distances using the latest video conferencing.

Following piloting in a mock shop Standard Chartered have started to roll out the concept. Branches in Hong Kong have already opened with more being constructed in India, Korea, China, Nigeria, Kenya, Singapore and Malaysia. By the end of 2011 there will be over 200 new model branches in around 18 countries.

STANDARD CHARTERED BANK

Design Company_**Allen International**

Location_Pan Asia/Global   Designer_Nic Preece   basic info

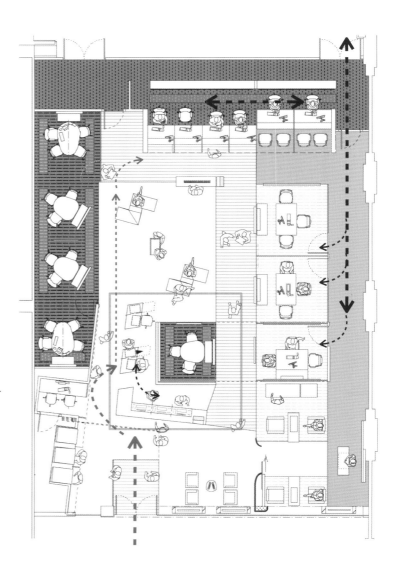

渣打银行网点遍布88个国家,在北亚和东南亚国家、非洲、阿联酋和印度次大陆拥有39家零售分行。

渣打银行邀请艾伦国际设计公司通过可控的成本、材料质量和布局,发展一种全球性的分行标准来传递银行一贯的品牌承诺。当这个任务完成之后,艾伦国际还会编制一份手册来支撑和巩固它。

艾伦国际专注于高度的人际互动,创造了一种战略措施,并通过分行内部的数字技术使用过程中所传递的"吸引、建立、俘获"这样一种零售通讯渠道进行支撑。

银行设置了一种托管式的数字媒体屏幕,在这里工作人员可以与客户传递讯息,客户则可通过交互式终端对产品和服务资料进行浏览、打印和收发电子邮件。客户和职工还可以使用最新的视讯会议开展远距离互动。

在对模拟商店进行试验之后,渣打银行已经开始推广这种理念。香港的分行开门大吉之时,在印度、韩国、中国内地、尼日利亚、肯尼亚、新加坡和马来西亚、越来越多的分行处于构建之中。到2011年年末,将会在18个国家和地区涌现出200多家银行模型分行。

_Description

A bank that starts from an association of ten batik merchants from Batik market "Pasar Baru" at Bandung, West Java in 1908 named "Himpunan Soedara" (Himpunan means Association). This design started with the momentum of the bank's 100th anniversary. Its historical value that the designers adopted as a design concept when the designers are commissioned to design the cilent's Priority Banking area (Area of a bank that dedicated for premium customer). This Batik pattern and the associations of ten batik merchants is the designer's main concept.

The designers want the batik patterns to fill the space. This pattern has to be the interior element and not merely decoration. They use this pattern to guide customers into the lounge and also as a skin between the corridor and the private staff & personal banking area, in the form of patterns on the walls and ceiling. The batik pattern that they use is the kawung pattern, it is symmetrical, and the word kawung means palm seeds.

For the concept of the ten merchants associations is represented with ten petals which became one as the service station. Aside from being the centrepiece, it also serves as cabling line for lighting and for the LCD TV's structure placement.

BANK SAUDARA PRIORITY BANKING

Design Company_**id.TECTURE**

Area_450m²

Location Graha Energy, 2nd fl. Jl. Jend. Sudirman Lot 11a, SCBD, Jakarta   Designer_Rinto Wiharjo, Hartono Dharmadi

Kawung as Pattern

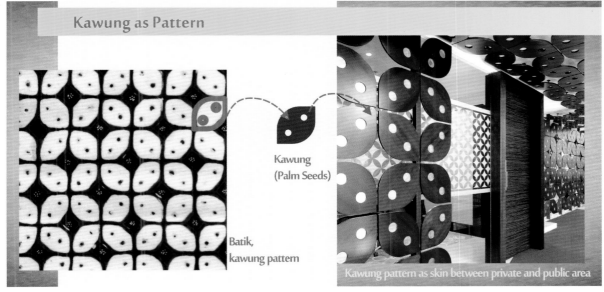

Batik, kawung pattern

Kawung (Palm Seeds)

Kawung pattern as skin between private and public area

索达拉优先理财银行起始于由10个蜡染商人组成的一个联盟，称做"Himpunan Soedara"，（Himpunan即联盟的意思），它形成于1908年的西爪哇岛万隆的一个叫"Pasar Baru"的蜡染市场。为了庆祝其100周年纪念日，才设计了该银行。当设计师们被委任来设计客户优先理财区域（致力于服务优质客户的银行区域）时，他们采用其历史价值来充当银行的设计理念——蜡染图案以及10个蜡染商人组成的联盟。

设计师们想用蜡染图案来填充空间。这种图案不单要有装饰效果，更要成为内部元素。设计师们将图案装饰在墙和天花板上，引导客户来到休息室，同时也将其作为走廊、私人服务人员以及个人业务区的分界。他们选用kawung图案，这种图案比较匀称，kawung这个词的意思为棕榈种子。

10片花瓣代表10个商人联盟，它们合十为一形成服务站。除了作为核心装饰物之外，它还起到了为照明设备和液晶显示屏的结构布局提供电缆的作用。

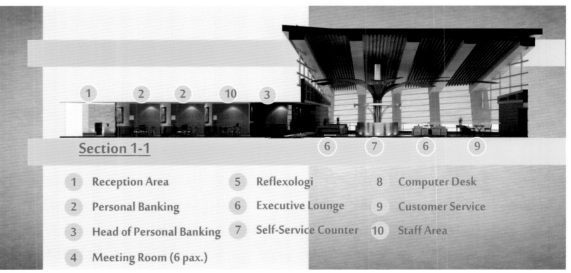

Section 1-1

1. Reception Area
2. Personal Banking
3. Head of Personal Banking
4. Meeting Room (6 pax.)
5. Reflexologi
6. Executive Lounge
7. Self-Service Counter
8. Computer Desk
9. Customer Service
10. Staff Area

Kawung as Geometrical Shape

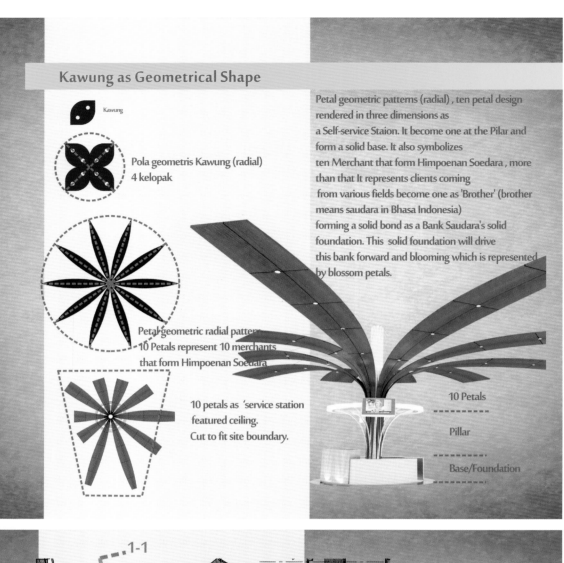

Petal geometric patterns (radial), ten petal design rendered in three dimensions as a Self-service Staion. It become one at the Pilar and form a solid base. It also symbolizes ten Merchant that form Himpoenan Soedara, more than that It represents clients coming from various fields become one as 'Brother' (brother means saudara in Bhasa Indonesia) forming a solid bond as a Bank Saudara's solid foundation. This solid foundation will drive this bank forward and blooming which is represented by blossom petals.

- Pola geometris Kawung (radial) 4 kelopak
- Petal geometric radial pattern 10 Petals represent 10 merchants that form Himpoenan Soedara
- 10 petals as 'service station featured ceiling. Cut to fit site boundary.
- 10 Petals
- Pillar
- Base/Foundation

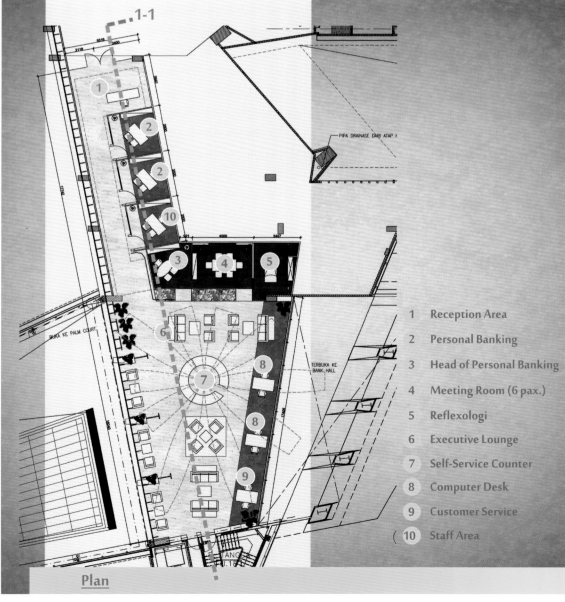

1. Reception Area
2. Personal Banking
3. Head of Personal Banking
4. Meeting Room (6 pax.)
5. Reflexologi
6. Executive Lounge
7. Self-Service Counter
8. Computer Desk
9. Customer Service
10. Staff Area

Plan

BCC BANK

Design Company_**Kuadra Studio**

| basic info | Location_Madonna dell'Olmo | Area_120m² |

_Description

The brief of this project is to convert an existing 2-floor building to be the new branch of the BCC. The floor area is 120 square meters and has an irregular shape. The original flooring will be conserved; thus, the project makes use of innovative solutions to create eye-catching visual effects. The drop ceiling is composed of large suspended panels in wood, which cast a pattern of light and shadow on the surfaces below. Utilities and cabling are fitted behind the ceiling. The calm, cool atmosphere of the office space is characterised by the contrast between the light-coloured areas of the walls and counter and the dark surfaces of the countertops. This effect highlights the wood panels forming the ceiling. The black wall surfaces made of panels with vertical bars open up to reveal storage cabinets.

The existing spiral staircase has been masked by a large cylinder, composed of tubes of varying diameters. These are mounted vertically and are irregularly spaced. The large white counter is divided into windows for the tellers by vertical glass "blades", to create a certain degree of privacy. Two large offices are partitioned off by glass walls.

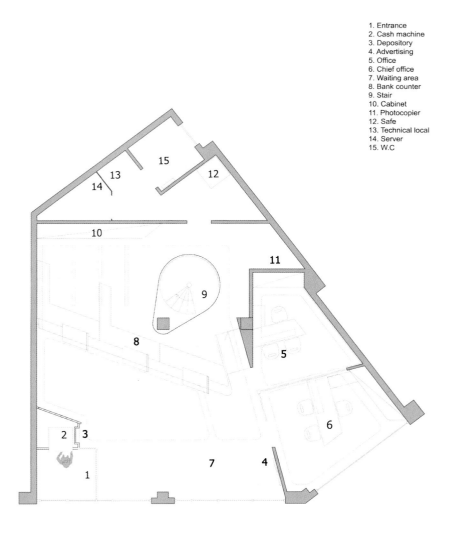

1. Entrance
2. Cash machine
3. Depository
4. Advertising
5. Office
6. Chief office
7. Waiting area
8. Bank counter
9. Stair
10. Cabinet
11. Photocopier
12. Safe
13. Technical local
14. Server
15. W.C

　　项目的目的是将现有的一幢2层大楼转变成BCC银行的一个新的分支。楼面面积是120平方米,并有着不规则的形状。原有地板将会被保留下来;因此,该项目利用创新的解决方案来创建醒目的视觉效果。吊顶是由大量悬浮板木材组成的,其光影图案会映在下面的地面上。工具和电缆安装在吊灯后。浅色的墙壁和柜台台面与深色工作台面的反差刻画了办公空间个性、安静的氛围,凸显了木工板制成的天花板。黑墙表面镶嵌着垂直条,打开便能看到储存柜。

　　现有的螺旋楼梯被一个由不同直径的管子组成的大圆筒所覆盖。它们都是不规则垂直安装的。出纳员的白色大柜台被垂直玻璃片分成若干个窗口,来创建一定程度上的隐私性。两个大型办公室用玻璃墙隔开。

_Description

Amsterdam firm Sander Architecten completed the Square of the office for Rabobank Nederland in Utrecht. Rabobank selected Sander Architecten out of a group of twenty to create and supervise the execution of the entire interior design (56,000 m²), including the twenty five-storey building. As the office interior is being redefined by the introduction of new methods of working, interior architecture is facing new challenges. In today's work environment, the emphasis is on cooperation in teams and group dynamics, people go to the office for the social aspect more than anything else. To realize this ambition, the designers view the building as a modern city. After all, the city is where individual freedom and spontaneous interaction are all-important.

The effectiveness of this concept is visible on the Square, located at the plinth of the new office building. Employees and visitors work, eat, read, and meet one another in a diverse landscape. The "buildings", separate spaces with different functions, join up with the uncluttered grid of skylights and slim columns. The new style of working is based on freedom, trust and taking responsibility. In the client's view, its employees are all entrepreneurs, responsible for their own performance in an environment free of fixed rules, fixed times and fixed locations. The work spaces are tailored to specific activities: multi-person meetings, face-to-face meetings or a place to write a report with maximum concentration. Each activity has its own space.

In nature routes are formed naturally; people intuitively find their way. Architect Ellen Sander was seeking that naturalness that "flow". The busiest routes automatically formed around the cores with the lifts and staircases, beyond which more peaceful zones naturally emerged. Moreover, the psychological concept of "flow", the moment when need, desire and ability come together, connects the employee's sense of happiness with an optimum result for the employer. The guiding principle for the interior design therefore became "form follows flow". To enable flows vertical partitions were avoided so that the horizon would always be visible. "The office is my world and the world is my office".

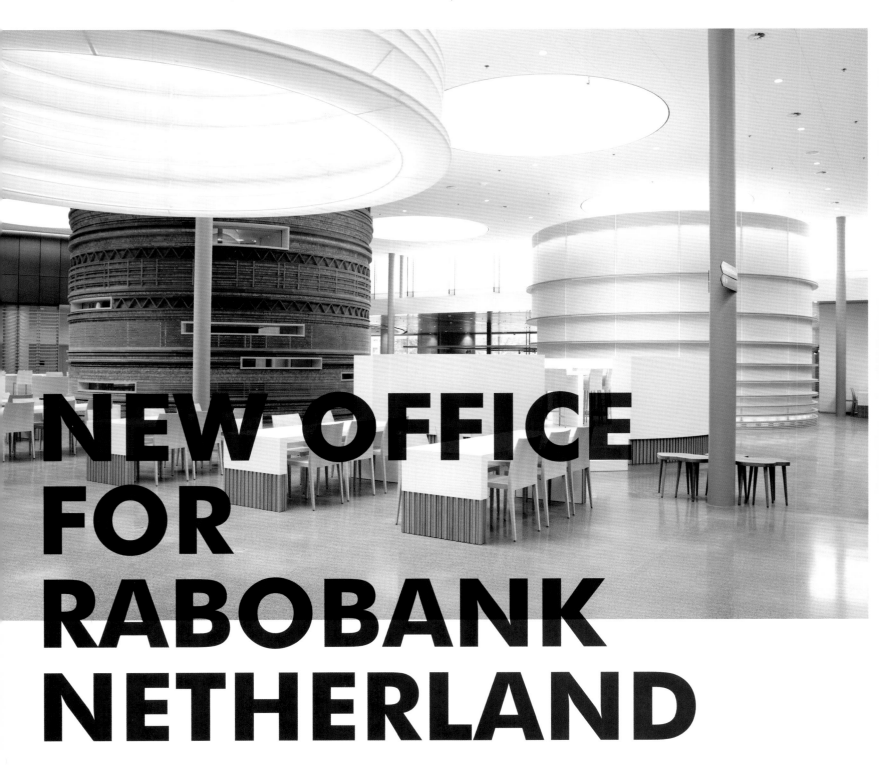

NEW OFFICE FOR RABOBANK NETHERLAND

Design Company_**Sander Architecten**

Photographer: Alexander van Berge, Ray Edgar Area: 56,000m² Location: Utrecht, Netherland basic info

阿姆斯特丹的Sander Architecten 公司在乌特勒支为荷兰合作银行完成了建造正方形办公楼的任务。荷兰合作银行在20家公司中选择了Sander Architecten公司来进行室内设计（共56000平方米），并监督整个设计的执行情况，包括这个25层高的建筑。由于引进了新的工作方法，人们对办公室内部设计有了新的定义，室内建筑也面临着新的挑战。如今的工作环境中，人们强调团队合作和集体动态，人们去办公室主要是为了一些社会原因，而并无其他。为了实现这个目标，设计师把大楼看成一座现代城市。毕竟一个有个人自由，可以随意交流互动的城市是非常重要的。

你可以从这个位于新办公大楼的底座的广场上直观地看到这个概念的效果。员工和访客可以在不同的景观中工作、吃东西、阅读、与朋友会面。这个建筑将空间分成功能各不相同的小块，与整齐的网格天窗和小柱子相连接。这种新的工作模式建立在自由、信任和承担责任的基础上。就客户看来，银行员工都是企业家，他们要在没有固定规则、固定时间、固定地点的环境中对自己的表现负责。工作空间的调整是为了适应某些特定的活动：多人会议、面对面会议或是一个高度集中精神写报告的地方。每个活动都有其自己的空间。

事实上，路线是自然形成的；人们凭着直觉找到路。建筑师Ellen Sander 寻求会流动的自然。以电梯和楼梯为中心自动地形成了最繁忙的路线，在周边自然而然地出现了越来越多的安静区。而且，"流动"这一心理概念，即当需要，渴望和能力汇聚到一起时，员工就既能获得幸福感，还能获得老板所希望的最佳成绩。因此室内设计的指导原则就是"形式跟随流动"。避免出现流动的纵向分区，确保可以一直看见外面的景观。"办公室是我的世界，我的世界就是办公室"。

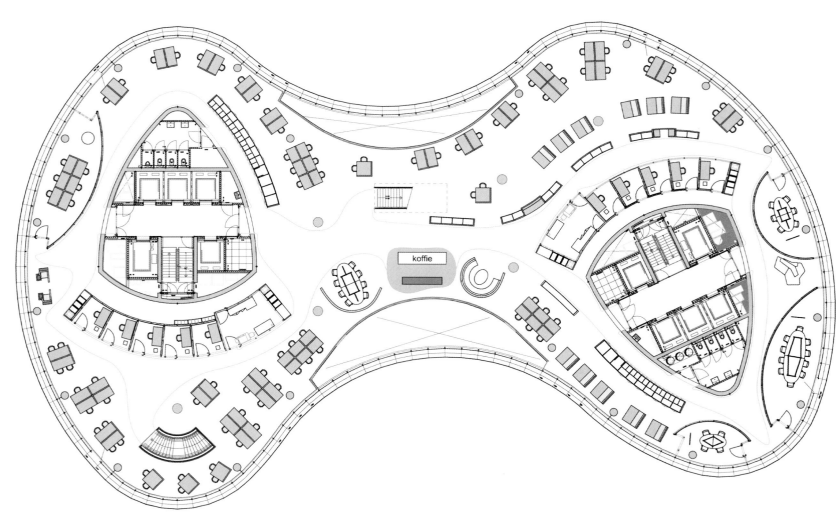

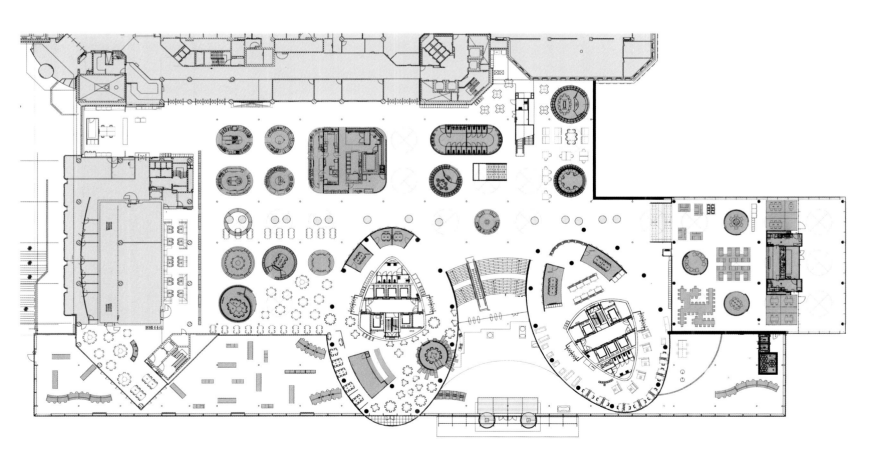

BNP PARIBAS BANK

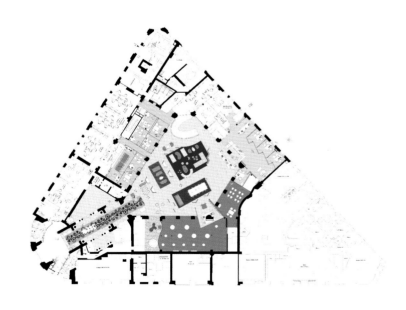

Design Company_**Zoevox Architects**

_Description

A user-friendly place set around a vast lounge dedicated to "active waiting" or leisure time. The 2 Opera defies the usual aesthetic codes of the banking world. This laboratory-shop where innovation and new customs are encouraged allows a new vision of the banking relationship. The transparent walls allow for an understanding of the building's broad volume while the specific harmonious or contrasting color schemes make the different living areas easily identifiable one from the other. Furthermore, Zoevox has chosen a few historical references (honeycombed ceiling, sophisticated mirror work) to make this unusual space an elegant link between past and future. Over almost 1,000 square meters, living areas with contrasting colors and furniture are designed to enhance the relationship between the bank and its clients, while stimulating a pro-active attitude in the latter. At 2 Opera, a creative scenography leads the client to discover the various services offered by Paribas in a playful and interactive way. He can gather information on the stock-exchange on his own or with the help of an advisor. Private rooms are dedicated to more formal appointments. An additional temporary exhibition area as well as another exclusively dedicated to children completes this new offering entirely devoted to an innovative client experience, in a surrounding destined to encompass true French elegance. On both sides of the entrance hall, the particularly unusual vegetal design of the ATMs set the spirit of the self service area (cash withdrawals, change, cash deposits etc.). In the lock chamber towards the welcome area, two screens inserted into the walls and placed behind mirrors broadcast information in a most poetic way. The welcome desk of 2 Opera is a long central table around which advisors are at the disposal of clients and prospects for guidance and directions. On the left-hand side is the exhibition area bordered by red walls, and on the right-hand side lies the monumental staircase leading to the management's offices.

Photographer_Veronique Mati   Location_Paris, France   Designer_Fabrice Ausset

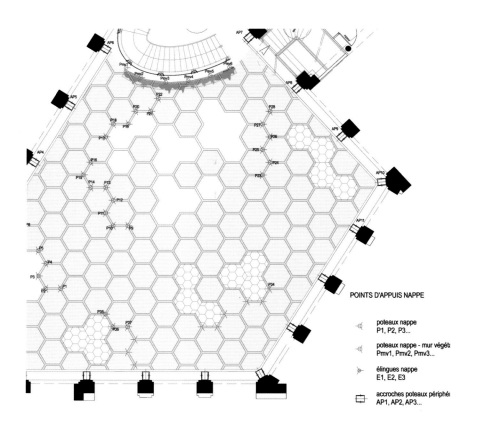

POINTS D'APPUIS NAPPE

- poteaux nappe
  P1, P2, P3...
- poteaux nappe - mur végét:
  Pmv1, Pmv2, Pmv3...
- élingues nappe
  E1, E2, E3
- accroches poteaux périphé
  AP1, AP2, AP3...

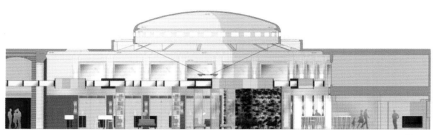

sections

法国巴黎银行坐落于一个人性化的位置，其周围是人们耐心等待或娱乐休闲的巨大休息厅。两家歌剧院对银行界一贯的审美标准发起了挑战。这一提倡创新和吸纳新风俗的实验室风格的建筑物为银行关系开辟了一种前所未有的视角。透明的幕墙保证了银行内部广阔的空间，同时这种特定的和谐、鲜明的配色方案，使得各个生活区之间产生明显的区别。此外，Zoevox采用了复古的风格（蜂巢天顶和精致的镜面），为这个独一无二的空间创造出过去与未来间典雅的联系。大约1000平方米的生活区被赋予鲜明的色彩和精美的家具，增强了银行与客户之间的关系，同时激发了客户积极主动的态度。两家歌剧院中，一种充满创意的透视图以诙谐而又交互的方式诱惑着客户去探索巴黎银行提供的各种服务。顾客可以自己或者在客服的帮助下收集股票交易信息。私人房间更倾向于正式的会议。此外，室内还设有一块临时展区和小孩体验服务专区，致力于为客户提供一种全新的体验，全然展示了纯正的法国式典雅风格。在银行大堂的两边，植物造型的自动取款机非比寻常，展示了自助服务区（取款、兑换、存款等）的特性。欢迎区正对面的私人房间的幕墙上，两扇屏幕嵌入到镜子后面，以极其诗意的方式播报信息。两家歌剧院的欢迎台由一张长方形中环桌子组成，客服耐心等待在桌子周围，随时为客户提供引导和咨询工作。歌剧院的左边，展区依靠着红色的墙；歌剧院的右边是一座巨大的阶梯，直通管理层的办公室。

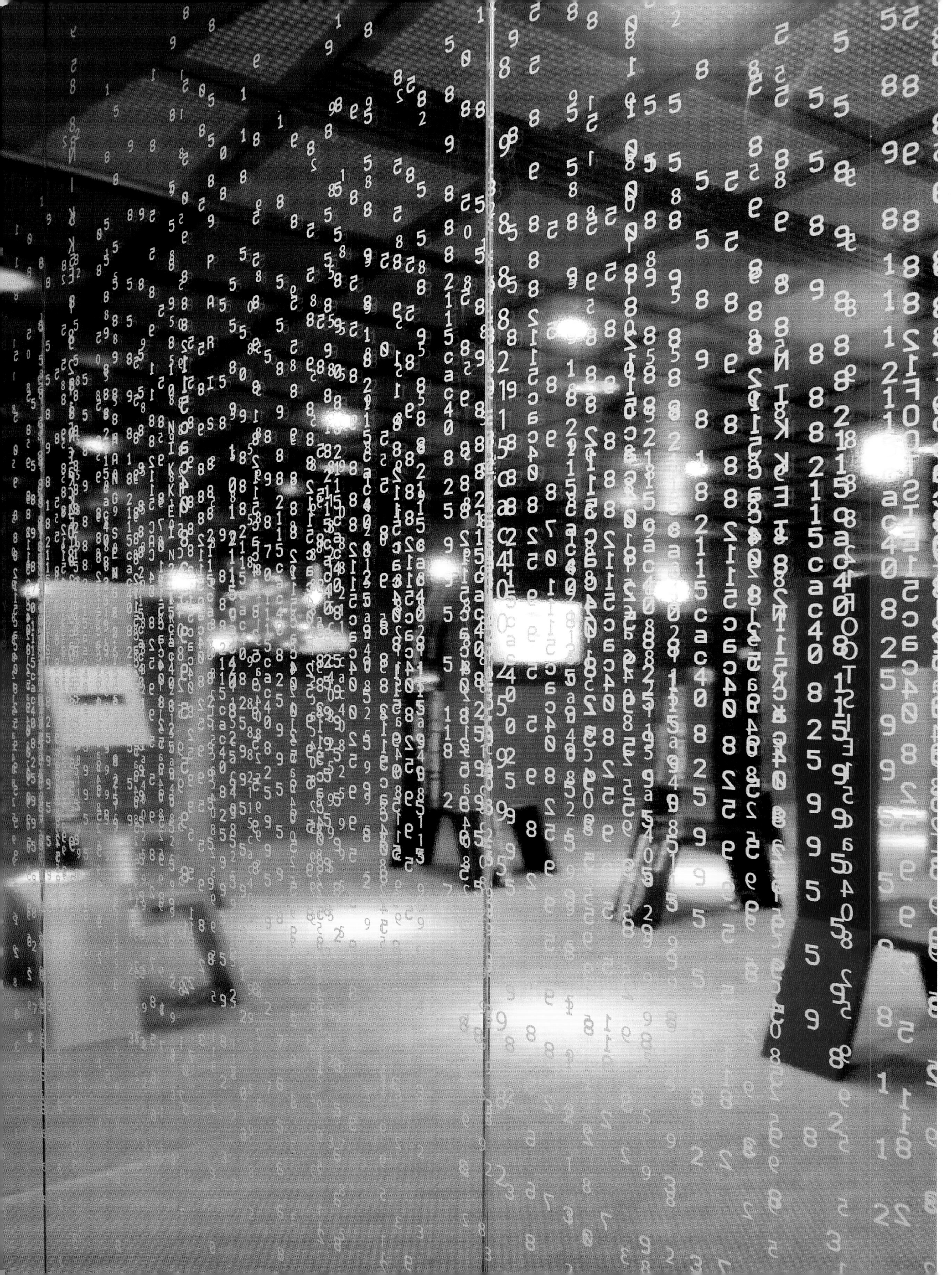

_Description

Allen international was commissioned to evolve the brand identity and retail branch design of the respected Emirati brand: Abu Dhabi Islamic Bank.

Building upon the brand's reputation for traditional Islamic financial services allen international's strategic goal was to make it much more relevant to a younger market and also to a much wider global non-Islamic customer base. To do this they subtly evolved the brand identity to reflect the transparency of the offer through its simple shape and rendering. The Brand's core DNA of Simplicity, Transparency, Partnership and Community established by allen international was the starting point for the retail branch design.

Taking the traditional souk marketplace stalls as the inspiration for this central area, allen international designed a flexible display and digital merchandise system. The teller zone was purposely recessed off the branch marketplace so it would not become a dominant feature of the retail offer preventing the branch from feeling purely transactional. Tactically positioned digital screens were sited in the branch as part of its Attract, Build, Capture (ABC) Digital communications strategy to create the dynamic delivery of promotional and brand messages to customers. allen international used this new branch design as the basis of the separate "Dana" ladies banking branches. By defining a unique colour scheme, brand pattern and materials a more feminine space was developed. Finally, the affluent segment "Diamond Banking" was developed. With its own concierge desk, exclusive lounge, prayer spaces and discreet meeting offices, a rich, exclusive design was created for this segment.

ABU DHABI ISLAMIC BANK

Design Company_**Allen International**

Location_Abu Dhabi   Designer_Richard Benson   basic info

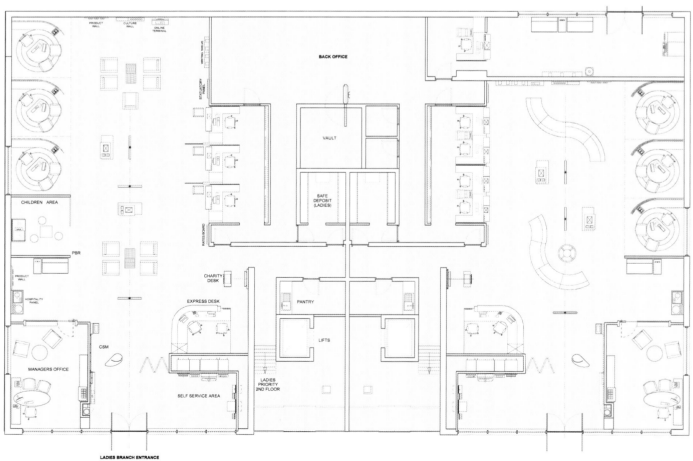

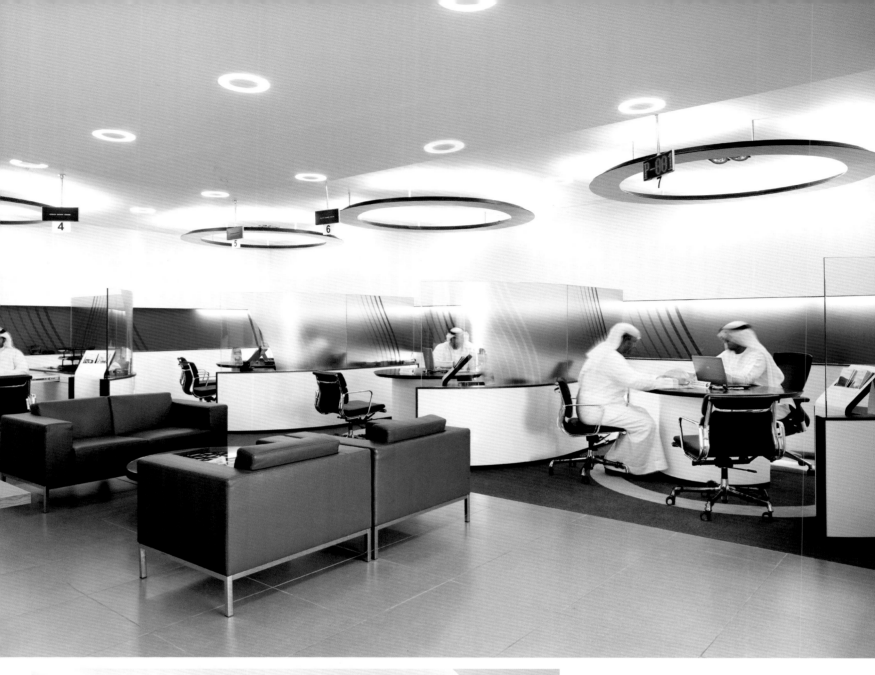

艾伦国际设计公司受阿布扎比伊斯兰银行的委托，在其原有的品牌形象基础上，升级其品牌形象及其在阿联酋的分行。

由于阿布扎比伊斯兰银行以其传统的伊斯兰金融服务而闻名，所以艾伦国际设计公司的策略是使其能更加贴合适应年轻市场，并把客源目标放大到全球性的非伊斯兰教的客户。因此，艾伦国际设计公司建筑设计的银行大厦以简洁、大方的建筑外观和抹灰效果将策略透明公开化，巧妙地将其品牌形象创新升级。阿布扎比伊斯兰银行的核心理念是"简洁、透明、合作、共享"，艾伦国际建筑设计则是这种理念宣传推广的起点。

中心区域的设计灵感来源于传统的市场摊位图，艾伦国际设计了灵活的陈列布局方式以及数字化的商品系统。特意将出纳员区域布置得远离中心区域，防止其喧宾夺主，给客户留下银行是一个纯粹金钱交易的场所的形象。将数字显示屏装在大厦室内，这种巧妙的做法也是吸引、构建、房获数字通信战略之一，把银行宣传推广活动信息以及银行品牌信息以动态方式传递给客户。同时，艾伦国际还将这种新颖的建筑设计风格作为银行单独隔离出来的"德纳女士"专区的建筑的基本模板。再通过独特的色系、品牌模式以及建筑材料等，成功地打造出一个更为女性化的空间。最终，华美精致的砖石银行完美收工。它拥有礼宾部柜台、专属休息室、祷告室和素雅简洁的会议室，一个丰富而独特的建筑设计成功地打造出了这个专区。

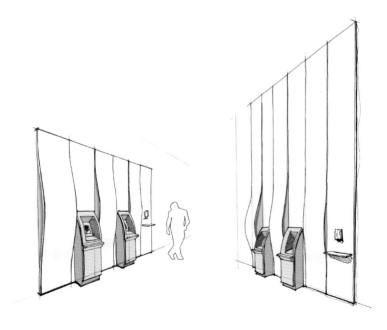

_Description

A brand new concept was developed for one of the most important Banks in Perú. Contemporary work requires new adjustable spaces, which were the center of the research made by the design team.
Based on that, the project tries to optimize the system, interaction and work team: every working spot was integrated to each other so that a central multi-proposal space would appear.
The intention at BCP was to reduce the occupation ratio without neglecting the quality of the office environment. The concept of lighting includes management systems that synchronized artificial and natural light at the same time to create a comfortable atmosphere. The design bet on inner perspectives to gain visual range and amplitude. The project is supposed to recreate the sensation of an Urban Café in an office context.

BANCO DE CRÉDITO DEL PERÚ

Design Company_**Architect José Orrego**

Photographer_**Juan Solano**

Materials glass, wood, big-sized graphics   Location CRONOS Building, Lima, Perú   Designer José Orrego – Metropolis   basic info

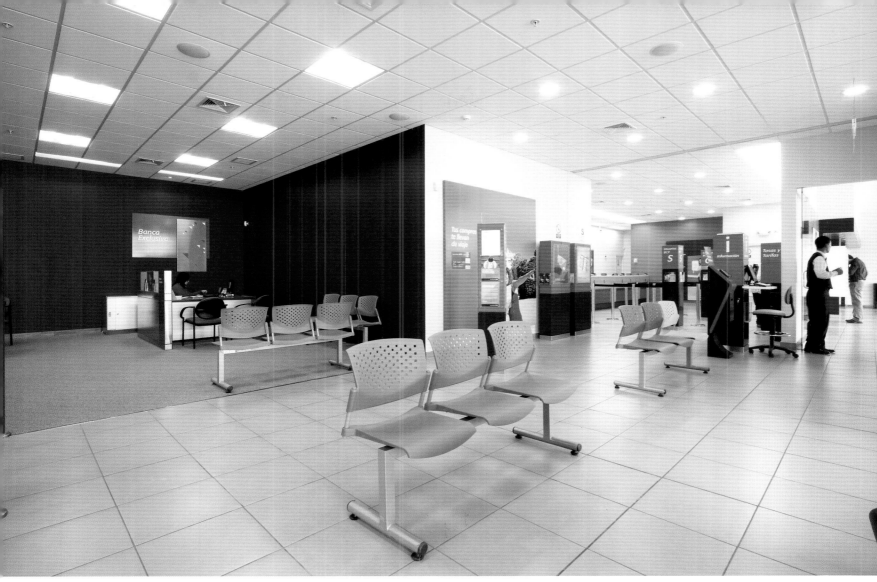

  秘鲁最重要的银行之一，秘鲁信贷银行最近研发了一种新的概念。目前的工作要求有新的可调整空间，这正是设计团队研究的核心。

  基于此理念，本案试图优化体制、互动及工作团队：每个工作岗位都应该相互协调，这样才能形成一个中心多功能空间。

  秘鲁信贷银行的目的是在不影响工作环境质量的情况下，减少空间占用率。灯光处理上，人造灯光与自然光相结合，共同打造舒适的氛围。该建筑的室内设计成功地延伸了空间感。设计试图在办公环境中重现城市咖啡厅般的休闲氛围。

BANCOLOMBIA

Design Company_Allen International

Bancolombia is the largest bank in Colombia with close to 900 branches there as well as a presence in El Salvador, Spain, Panama, Puerto Rico and the USA. To maintain its position as the number one financial services provider in Colombia, Bancolombia appointed allen international to develop its branch environments to enhance the customer experience and reflect the bank's brand values, leading to increasing revenues and profitability.

Allen international developed a retail design strategy that centered on reconnecting Bancolombia with the people and communities of Colombia. The concept of "Hub Colombia" was created as the centrepiece of the branch, connecting through pictures, words, films and music, the wide variety of regional people and their cities, towns and villages. Customers are invited to contribute their thoughts, images and ideas to an ever-growing digital database that can be viewed in branch or via the Internet. The collected images, sounds and films are an integral part of the branch design. These provide a sense of ownership and pride from customers and staff alike. Bancolombia has now implemented pilot branches in three major cities – Bogotá, Medellin and Barranquilla (7 pilots in all).

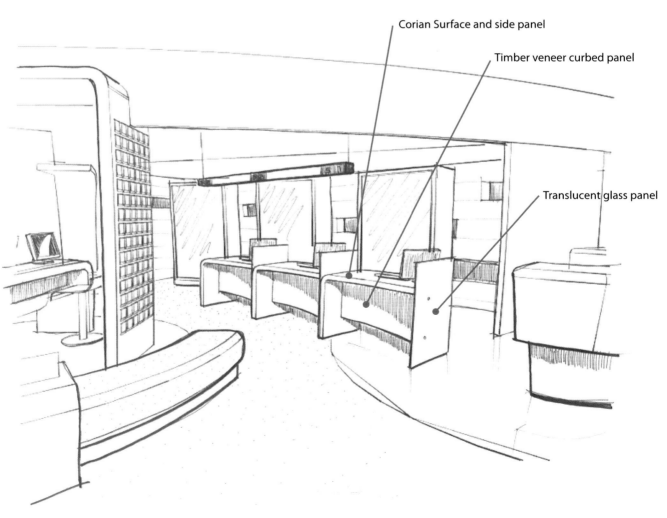

Corian Surface and side panel

Timber veneer curbed panel

Translucent glass panel

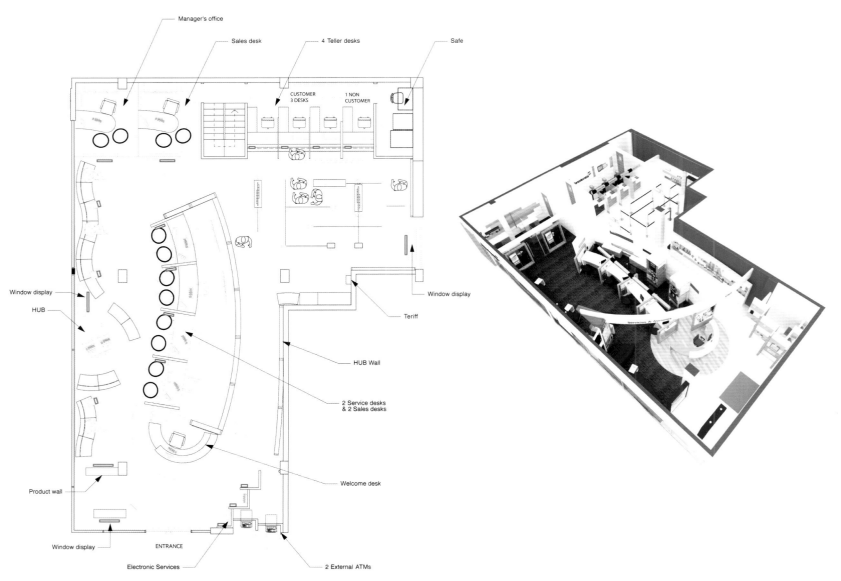

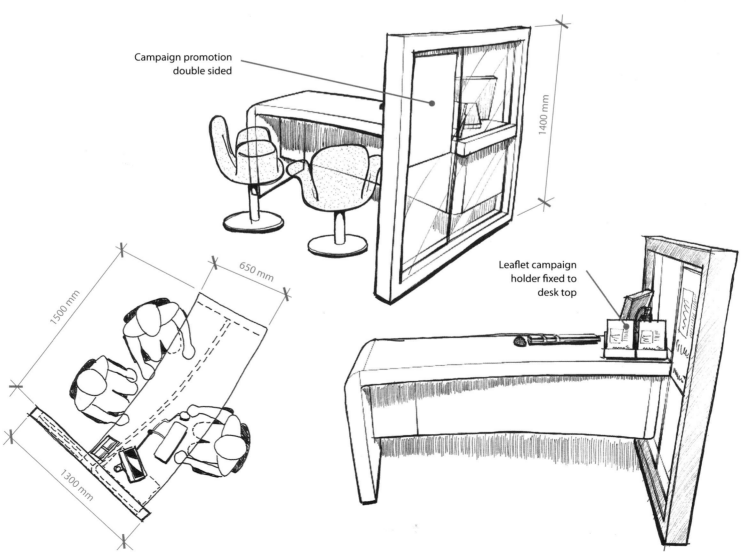

哥伦比亚银行是哥伦比亚最大的一家银行，拥有分行近900家，并且在萨尔瓦多、西班牙、巴拿马、波多黎各以及美国都有其办公处。为了保持其在哥伦比亚金融行业中第一把交椅的地位，哥伦比亚银行委派艾伦国际设计公司升级和重新装潢其中一些分行的建筑环境，从而改善和提高客户体验以及反映其品牌价值，达到最终的提高收入利润的目的。

艾伦国际为其制定了一个设计策略，就是将哥伦比亚银行再次融入到哥伦比亚的人民和社群团体中。"哥伦比亚的中心"这一概念是分行建筑设计的核心，而再次融入是通过图片、文字、宣传片、音乐、各个不同地区的人以及他们所在的城市或乡镇这些元素达到的。客户被邀请针对能在银行内浏览或网上查阅的日益增长的数字数据库发表他们的看法和观点，这些收集来的图像、音频、视频是分行设计的重要组成部分。这能够给客户以及员工一种所有权，一份自豪感。哥伦比亚银行已完成了设在以下三个主要城市试点分行的银行大厦建筑：波哥大、麦德林和巴兰基亚（共7个试点分行）。

I-BANK

Design Company_**Allen International**

_Description

The National Bank of Greece came to allen international with the challenge of designing a space where electronic banking is for everyone, not just the technically savvy. allen international's extensive knowledge of the customer experience and Attract, Build, Capture, (ABC) Digital solution was the backbone of ensuring NBG took itself, and its customers into the future of banking.
The National Bank of Greece has unveiled a pioneering bank concept store in The Mall, Athens, which uniquely combines contemporary architecture with the latest technologies. i-bank is a modern multifunctional space designed by allen international for NBG's alternative channel sub-brand. Visitors are entertained with games, can enjoy a relaxed drink and use all i-bank services. The new branch combines technological innovation and contemporary design with ecological principles. i-bank is a technologically innovative, ecologically responsible, and customer friendly space that changes peoples' perception of a bank.
"i-bank" is a multi-purpose electronic banking environment that delivers services via Internet Banking, Mobile Banking, Phone Banking, ATM and Automated Payment Systems. The visitor, either alone or with the help of a specialist from i-bank store team, can explore a new quick and easy execution of daily banking transactions. i-bank is a place of entertainment, training and an exchange of views on technology, modern banking services and the environment.

Location_Greece   Designer_Richard Benson   basic info

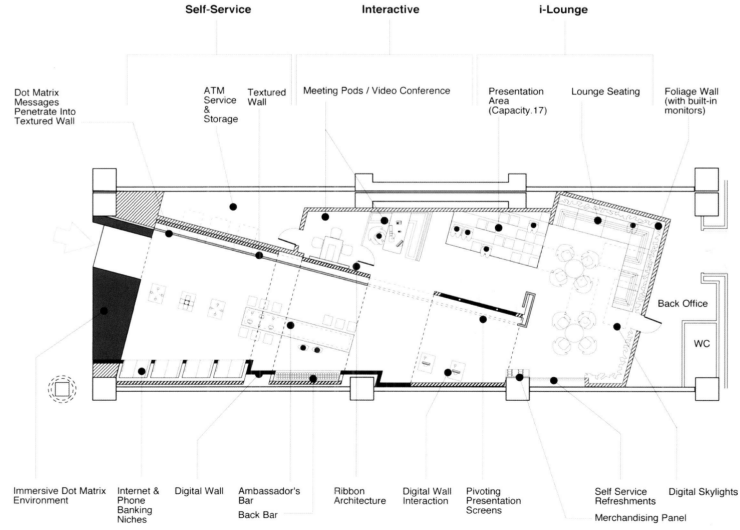

　　希腊国家银行带给艾伦国际一个挑战，那就是设计一个空间，在那里电子银行业为每个人服务，而非单单服务于那些懂技术的人。艾伦国际在客户经验、吸引力创造、建筑、占领市场、数字解决方法方面知识渊博，这些优势是确保希腊国家银行走向银行业前沿、吸引更多客户的关键因素。

　　希腊国家银行在雅典商场内开设了第一家开拓性的银行概念店，它将当代建筑和最新技术相结合。网上银行是艾伦国际设计的一个现代多功能空间，它是希腊国家银行的子品牌，是它的替代渠道。来访者可以玩游戏，自我娱乐；也可以喝点饮料，放松一下，享受所有网上银行的服务。这个新的分行把技术革新、当代设计与生态原则有机地结合起来，是一个技术创新、对生态环境负责、让客户感到自在的空间，这里会改变人们对银行的看法。

　　网上银行是一家多功能电子银行，它通过网上银行、手机银行、电话银行、自动取款机和自动支付系统来提供服务。来访者无论是自己办理业务，还是在网上银行专家团队成员的协助下办理业务，都可以快速找到一种新颖、快捷的业务办理方式。网上银行是一个可以娱乐、培训，就科技、现代银行服务和环境交换各自看法的地方。

Description

Airbank is a new kind of bank store which aims to be noticed and admired, where people can experience in the name of simplicity, transparency and elegantly, with natural and intuitive control, where all the rules have to be understood by all.

The customer as protagonist of a new bank trip where he must feel that's the unique bank he wants, where receiving the services he desires, where feeling released always with the consciousness that the bank is doing for him. Working on the worldwide methodology of Physical-brand-design, Crea International team approached the project in a very organized and structured way. An open service model with no constraints typical of traditional banking models : here people are invited to take a seat and are free to browse the bank offer and get the bank attendees support if needed. The result of our project is a totally new banking framework focused on the relationship and not the transaction: Crea believes the overall concept conveys a new perception of doing banking and will get people to adhere to the claim of airbank:" this is the bank I like ".

Aiming at simplifying the complex world of banking, using modern technologies, following Air bank's values of simplicity, transparency, environmental friendly, the result is to offer the world an extraordinary bank. All these features visible even in the simple logo, clean and sharp in a mixture of green and black colors giving freshness, vitality and the sense of renaissance and regeneration. The inspiration came from Agora landscape, in ancient Greek an open "place of assembly", to create an opened bank, where staff and customer can stay side by side, keeping out from the key points of traditional service models.

To do this, Crea's team has conceived an open squared place with just a multifunctional central area to make the space fruition free. All surrounded by an amphitheater with a stair seating system looking towards the center, where the customers can choose the seating on different levels according to the activity they have to do. Closed and private spaces are reduced in order to avoid any connection with the old world of "bank offices".

Finally a bank speaks a new language, setting the mark of a different bank which does not intend to astonish and confuse, but innovate to create value.

AIR BANK

Design Company_ **Crea International**

Designer_ Andrea Borsetto   Concept supervisor_ Massimo Fabbro   Architect_ Armando Loreti

Strategy director_Viviana Rigolli   Lead graphic designer_Sonia Micheli   Graphic designer_Gilberto Vizzini

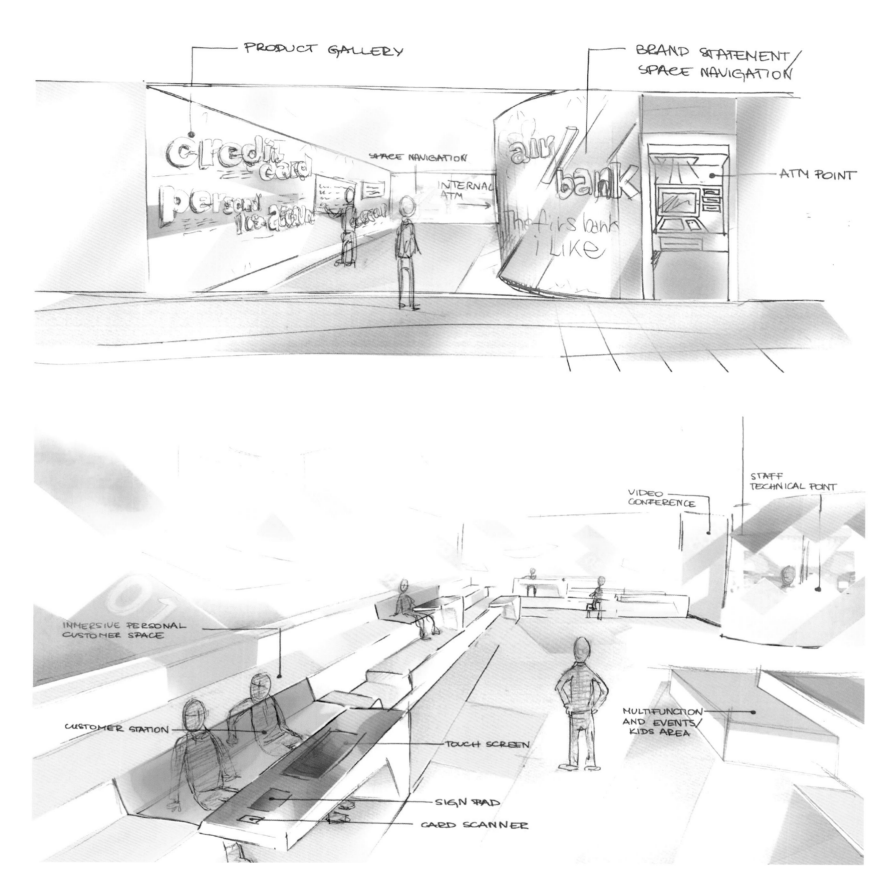

空气银行是一个旨在被发现和被推崇的新型银行，在那里人们可以通过自然、直观的控制，了解所有的规则来体验简单、透明化和优雅。

客户作为一个新的银行之旅的主角，他会感觉到那是他想要的独特的银行，能够得到他想到的服务，在接受服务时感到放松愉快。为了致力于全球体育品牌设计学，Crea国际团队赋予此项目组织性和结构性。一个不受典型传统银行限制的开放服务模式：在这里，人们可以随便休息，浏览银行报价，如果有需要的话还可以获得银行工作人员的帮助。该项目将会成就一个全新的银行结构、重视关系而非交易：Crea国际团队认为此项目的整体概念会传达出一种新的银行业的观点，并让人们坚信空气银行的追求："这就是我喜欢的银行"。

旨在利用现代技术简化复杂的银行业，遵循空气银行简单、透明、环保价值观，最终成就了一个杰出的银行模式。所有可见的特点，即便是在简单的标志上的，都是绿色和黑色的综合，干净、强烈，给人一种清新感、活力感和文艺复兴的再生感。此灵感来源于城市广场的景观，在古希腊的一个开放的"集会"，创造一个开放的银行，员工和客户可以并肩站着，改变了传统的服务模式。

为了做到这一点，Crea国际团队设想将这个开放的广场设计成多功能区，使其变得更自由。周围为面向着中心的、有着许多台阶座位的露天广场所包围，顾客可以根据自己的活动选择坐的座位。为了避免与旧有的"银行办事处"雷同，将尽量减少私人、封闭的空间。

最后，银行将会说一种新的语言，标志着它是与众不同的，不会使人震惊和迷惑，却能创造新的价值。

EXTRABANCA

Design Company_**Crea International**

_Description

The projectual metaphor of this new retail design concept is represented by the bridge which is the distinctive architectural feature of the place. It is seen as a sinuous portal of wood which spaces out the overall surface, articulating the operational desks and welcoming customers to a new reassuring retail banking experience. A powerful dynamic sign decisively translates the sense of continuity of the project and breaks up at same time the most conformist retail banking design layouts.

From outside a peculiar stylistic role is played by the window system which is made of "canaletto" walnut wood reporting the backlit logo of Extrabanca, while the flag ones located in the interior enhance the visibility through the deep red colour of the logotype. The visual impact is furthermore enhanced by a sophisticated dynamic lighting system of Led RGB.

As for the finishings and the colour palette, an institutional choice through an elegant chromatic mood transfers inside the retail design concept of the corporate colours: red, white and barrel grey. Finally the white architectural box results to be scanned by an entrance portal distinguished by a deep lacquered red, pointing out the meaningful presence of the brand since the first step into the branch.

Once again Crea International succeeded in conceiving such an innovative retail banking design concept able to communicate through design a series of values related to the respect of listening, pride and customized service.

　　这个新的零售设计概念投影效果是通过该地具有特色的建筑特征的桥来表现的。它被看成是一个蜿蜒的木头入口，它将整个表面分隔开来，连接了操作桌子，欢迎客户体验令人安心的零售银行业务。一个强大的动力标记果断地表现了这个项目的连续性，同时打破了传统的零售银行设计布局。

　　从外面看，窗户扮演着一种特殊体裁的作用，它由卡纳莱托核桃木制成，反应出Extrabanca的背光标志，同时，位于室内的标示通过标识的深红色字体加深了可见度。高级的、动感的红绿蓝LED（发光二极管）灯照明系统则进一步增强了视觉效果。

　　关于最终的颜色搭配，通过一种优雅的彩色心情做出了一个制度选择，将内部设计概念转换成公司的三种颜色：红、白、桶灰色。最后白色建筑盒子被深红色油漆大门掩盖了，踏入这个分行的第一步，您就能感受到这个品牌的重要意义及其突出的存在感。

　　Crea国际又一次成功地构想出了这样一个创新性的零售银行设计概念，这个概念表示通过设计一系列与倾听、自豪和定制服务相关联的价值观，能传达出背后的意义。

DEUTSCHE BANK, THAILAND

Design Company_**Orbit Design Co., Ltd**

Description

Modern financial workplaces need to embody the same core messages that the whole industry does in the current market: openness, transparency and efficiency. The design for Deutsche Bank's Thailand HQ takes these principles to create open, flowing spaces, with line of sight connectivity between client hospitality areas, meeting rooms and work floors.

Touchdown areas allow for informal meetings and a welcoming environment for visiting staff, and a highly efficient floor plate maximizes the usable area. Key to the project was reflecting distinguished status of the bank, yet also using significant local influences that show their commitment and respect to Thai culture. Therefore, their office at the Athenee Towers is a beautiful convergence of Thai elements and modern touches.

The color palette of the interiors takes direct reference from the colors of Thai temples and palace roofs, dating back from the Sukhothai period. The colors yellow, orange and green are assigned to the three floors of the Deutsche Bank office and are picked up on the wall finishes throughout the working areas.

现代银行办公场所需要传递的核心信息为：开放、透明和高效，全行业概莫能外。德意志银行泰国总部的设计吸取了上述原则，创造出了开放和流动的空间，把客户接待区、会议室和工作区连接在一起。

接待处可以进行非正式聚会和用做欢迎访客，而高效的楼板可以最大限度地利用空间。项目的关键之处是反映出银行尊贵的地位，同时也通过使用重要的当地影响力来显示他们的承诺和对泰文化的尊重。因此，德意志银行在Athenee Towers的办公室是泰元素和现代格调的美丽结合。

银行内部颜色直接参考了泰国寺庙和素可泰王朝宫殿屋顶的色彩。三层的德意志银行办公楼使用了黄色、橘色和绿色三种颜色，办公场所墙面也刷有此三色。

_Description

The project concerns the redesign of a building in Donoratico (Livorno) located next to the head office of the bank "Banca di Credito Cooperativo di Castagneto Carducci" (also designed by architect Massimo Mariani in 2002).
The building is on two levels. From the functional point of view the ground floor houses the areas most strictly connected with the banking business like the big hall with counter and advisory services for customers. The branch management offices as well as the loan management and administrative offices are encapsulated inside a long slice of colored bureaus on the right of the entrance. A lot of other banking offices, safe deposits and services areas (vault, archives, etc...) are in the basement.
Inside the hall, upper the round waiting seat, there is a plasterboard ceiling randomly performed. Like a big drop, it comes out from the elongated corridor giving light to directional bureaus, service offices and common spaces.
A single graphic sign features all public spaces which are accessible to customers. All the furnishings and wooden wall are designed with vertical stripes in red, blue, yellow and green. In this way areas that in banks have traditionally been somewhat staid and "bleak" have been reinvented to give them an amusing and more optimistic look.

BANK IN DONORATICO

Design Company_ **Massimo Mariani studio**

Photographer_Alessandro Ciampi   Location_Livorno, Italy   Designer_Massimo Mariani   basic info

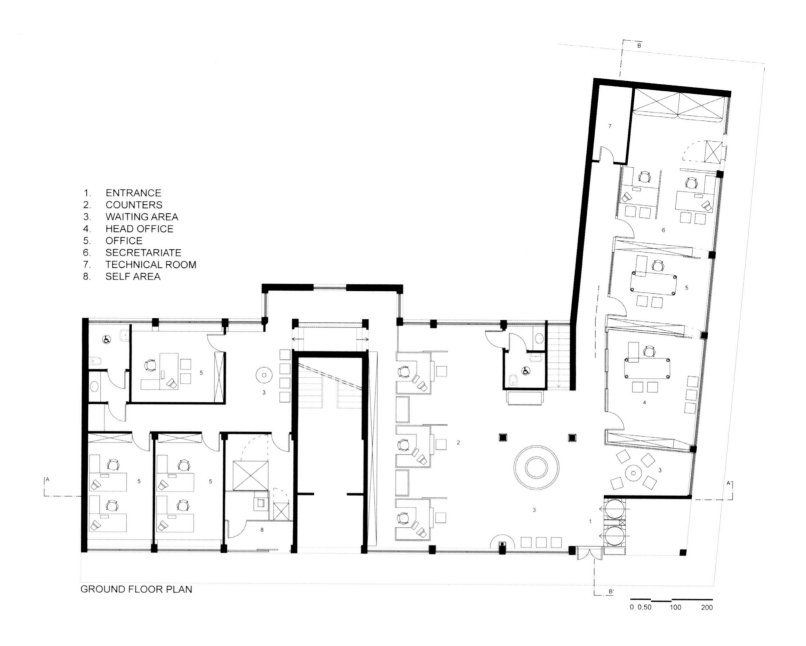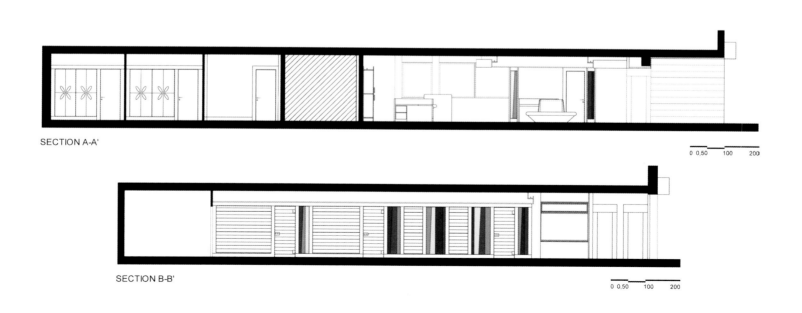

该项目是重新设计一栋位于意大利里窝那Donoratico地区的建筑，该建筑与"Banca di Credito Cooperativo di Castagneto Carducci"银行总部（2002年同样由Massimo Mariani设计完成）相邻。

建筑共有两层。从功能的角度来看，一层主要布置与银行业务相关的功能区，例如，带有柜台的大厅及客户咨询区等；而分行的管理办公室、借贷管理及行政办公室等布置在入口右侧长条形的彩色办公区内。其他一些办公室、保险库及服务区（贵重物品保管库、档案等）被布置在地下室。

大堂内，在等候席位的上方有一个石膏板制天花板。它就像一个大断层，从狭长的走廊中突然冲出来，为服务区及公共空间提供照明。

一个简单的标志凸显了给客户提供服务的所有公共空间。所有的家具和木墙都是由漆成蓝色、黄色、绿色的长形条木板组合而成的。通过这种处理，一扫银行给客户带来的古板、"萧条"的感觉，使整个空间变得更加活泼明朗。

_Description

As an established Oklahoma City bank sought to enter the competitive Arizona market the design team considered branding as a top priority in the design process in order to set MidFirst Bank apart from the local competition. Creating a comfortable, hospitable environment on a branch bank budget is always a challenge; however, it was achieved by using local vendors and materials to keep costs down while supporting the local community at the same time. Natural materials tend to vary greatly in their finish, which is why we used a hand selection process to ensure uniformity. These warm, natural materials are accented by rustic orange colors resulting in a masculine yet sophisticated and hospitable style. In the desert climate controlling heat gain while simultaneously achieving the look of light filled space always poses a problem. We were able to overcome these obstacles by adding overhangs to the south and west sides of the building and installing automatic Mecho shades on the windows. Also, glass-enclosed rooms were oriented inboard with circulation adjacent to the wall of windows and internal atrium, which maximizes access to natural light and enhances the sense of space as well. Amenities include private offices, visitor offices, video conference rooms, break rooms, mail room, and full service bank security which incorporate a vault and security monitoring into the design process.

MIDFIRST BANKING

Design Company_**Rottet Studio**

Location_Scottsdale, AZ | Designer_Lauren Rottet | basic info

Finish Plan

One Renaissance Furniture Plan

Floor Plan

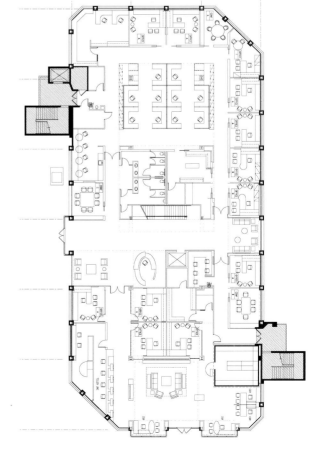

作为俄克拉荷马市一家既定的、旨在打入亚利桑那州激烈市场的银行，设计小组把银行品牌作为设计的首要任务，从而使MidFirst银行在当地的竞争环境中脱颖而出。要用建造分行的资金打造出一个舒适宜人的环境，是一个不小的挑战；然而，此设计通过利用当地供应商和原料降低了成本，同时，还支持了地方经济。自然材料在装饰时往往变化较大，因此我们通过人工筛选来确保它的均匀性。由于乡土橙色的强化效果，这些色彩柔和的自然材料呈现出阳刚却又精致宜人的风格。在沙漠环境下，控制吸热的同时达到充满光亮的空间效果是很困难的。但是我们在建筑物的西、南面各增加悬臂结构并在窗户安装自动百叶窗克服了这种障碍。此外，玻璃封闭的空间面向内部，环流与玻璃墙和室内的天井相毗邻，这样不仅使自然光的射入达到最大化，还增强了空间感。银行设施多种多样，包括私人办公室、来客办公室、视讯会议室、休息室、收发室以及把地下室和安全监控合并到设计程序的全方位服务的银行警卫室。

NATWEST

Design Company_**Dalziel and Pow**

basic info | Location_UK

_Description

Dalziel and Pow was appointed by Natwest to create a concept for the future of their retail banking channel, to establish a branch network to deliver "Helpful Banking", to embody the strategy to become the UK's most helpful and sustainable bank. They created a concept with a customer focused design, which recognizes the need for a truly unique banking experience that reconnects with the customer on an emotional level as well as aesthetics.

The concept has a zonal approach to enable the customer to decide how they transact, at their own pace. The customer can choose technology when appropriate, or has the option to interact with staff if preferred. The ambiance has a domestic quality, but with a line of professionalism and cleanliness, underlining the feeling of efficiency with a human face – the heart of "Helpful Banking".

国民西敏寺银行任命Dalziel and Pow设计公司为其创造未来零售银行业务渠道的理念，建立分支网络传递"便民银行"的理念，使其成为英国最便民的可持续发展银行。设计师创造了一个以客户为中心的设计理念，该理念满足了真正的、独特的银行经验的需要，重新实现了与客户在情感及审美上的对接。

这个概念有一个分区的方法，客户可以根据个人情况来决定如何交易。顾客可以在适当的时候选择高科技，随自己意愿与员工互动。银行有着家庭般的氛围，同时也具备一流的专业精神和整洁度，以人情味的方式强调效率是"便民银行"的核心。

UOB BANK

Design Company_**Silverfox Studios**

| basic info | Designer_Jojo Avelino, Dennis Sumampong, Manolito Mercado | Principals_Susan Heng & Patrick Waring |

_Description

The journey up to the Privilege Banking facilities on the 20th floor begins at the ground level Concierge arrival lounge. Silverfox Studios designed the 56-square-meter Concierge to have a boutique tailored interior, which has a sense of place next to the marina bay. The ceilings are tall and the colors evoke both the brightly lit maritime feeling of the water, trees and landscape outside the concierge facility.

The natural elements of Water and Sky are at the entrance and they form a pathway which leads the guest through a virtual forest experience, into the lounge and hospitality suites beyond. This carpeted area has a hand tufted sculpted bamboo forest floor design which is in light tonal earth shades complimenting the soft light hues of the adjoining lounge areas. The entrance area pathway is a combination of dappled light through the forest canopy and soothing hues of blue water.

Patrick states that "Bamboo is an Iconic Asian image as a symbol of sustainability and longevity for centuries. As UOB is an Asian bank, the values associated with the Bamboo forest it is ideal as the main ingredient for the conceptual setting out for this branch."

The guest is guided through a virtual bamboo forest, achieved by laminating photographic transparencies onto curved glass screens, which rise through a mirrored ceiling. The screens and the conceptualization of this space was visualized by Silverfox, however the realization of this idea was achieved by the acclaimed International Artist and Photographer Russel Wong.

Within the Privilege Banking area, Susan and Patrick designed the meeting and lounge areas as a series of floating layered panels upholstered in soft warm leather. These screen walls which Russel Wong was commissioned to depict through photography of Orchid flowers, are woven into the architectural setting out of walls and ceiling. They have integral light and evoke the shapes and natural petal designs of orchid flowers. They serve as a backdrop to large comfortable sofas and lounge chairs.

Silverfox Studios set out the space in a natural, organic a-symmetrical format which connects the views and creates flowing circulation routes through the lounge to the large Suites beyond. There are meeting, dining and lounge areas within each Suite and they each have a subtle difference in the design approach.

The "Super Suite" has a majestic view of the Marina Bay and with a direct link via the concept, evokes the luxury finishes and affluence of super yachts and the maritime connection.

Area_1,335m²   Photographer_Darrell Gan

到银行大厦第20楼银行专属服务区的旅程始于一楼大堂。银狐工作室把这个56平方米的大堂内设计成一个裁剪讲究的精品店,体现其紧挨在滨海湾的职场感。天花板很高,所运用的颜色不仅能让人们强烈地感受到弥漫着海洋气息的海水和绿树,还能感受到大堂外面的怡人景色。

海水和天空这些天然元素被设计布置在主要入口处,形成一条通道,仿佛客户要穿过一片的森林,才能到达大堂,然后享受到银行工作人员的盛情招待。这片铺地毯的区域设计运用了手工雕刻的竹林图案的地板,在轻快的色调下,衬托出与之相邻的休息区的柔和的色调。主入口处的通道是从铺满植被的顶棚透进来的斑驳的光线和碧水的舒缓色调的完美结合。

帕特里克指出,"在亚洲,几个世纪以来,竹子都是可持续发展和生生不息的象征。大华银行是一家亚洲银行,将其价值观与竹林联系在一起,作为银行的核心概念的主要特色是最理想不过的"。

客户被引导穿过这片虚拟的竹林,这竹林是由一些层层叠叠的竹林照片穿插放在一个圆弧形的玻璃屏上,再由镜片装饰的天花板折射而制造出来的视觉效果。玻璃屏和这块空间设计的概念是由银狐工作室设想出来的,而把这个概念变成实体的却是备受欢迎的国际艺术家和摄影师Russel Wong。

在银行专属服务区内,苏珊和帕特里克将会议室区域和休息室区域设计成流动分层板,在板上安置上柔软舒适的皮革。屏幕墙上的装饰是根据Russel Wong拍摄的兰花照片编织而成的,并安置在墙和天花板的表面上。他们用完整的光线照射出兰花的形状,甚至是花瓣。他们还将这作为舒适的大沙发和躺椅的布景。

银狐工作室将空间设计规划成自然、有机并且协调的布局,创造和连接的流动循环路线使人能透过休息室看到大套房之外更远的景物。每个大套房内都有会议室、餐厅和休息室,但是它们在设计上又有着微妙的差异。

超级套房结合了滨海湾的宏伟景色以及向别墅过渡的概念,激发设计师使用豪华的装饰、富裕的超级游艇和海洋之间的连接。

_Description

In an empty tenant space in a struggling neighborhood, OneCalifornia has filled a vacancy in an area with many empty storefronts. Boosting financial literacy is an important component of the bank's mission. Recognizing that banks can be intimidating to first-time users in this tight, urban setting, design lends to a place that is comfortable and inclusive. The layout allows movement to the teller and banker stations, while creating an intimate waiting area. Broad views to the street outside give a greater sense of security to the neighborhood and encourage other businesses to set up shop there.

ONECALIFORNIA BANK

Design Company_ **Mark Horton / Architecture**

Photographer_Ethan Kaplan Location_Oakland, CA Designer_Mark Horton, James Stamp basic info

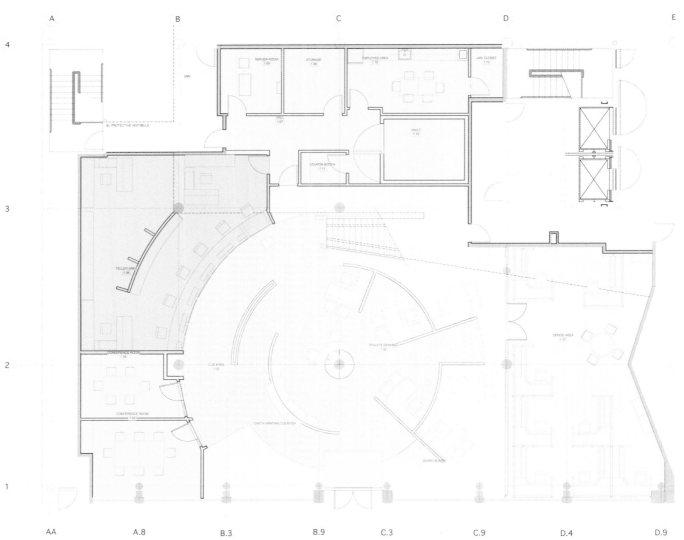

在一片不景气的空旷承租人空间中，OneCalifornia银行填补了拥有许多冷清店面的空白区。推动地区金融素养是该银行使命的重要组成部分。对于初次选配用户来说，银行严峻的城市背景着实有点吓人，因此，该银行的设计为它的地理位置提供了舒适又具有包容性的氛围。银行的布局在创造私人等候区的同时，保证了出纳和业者柜台的灵活运动。银行内部对外面街道有一种宽阔的视野，这样增强了街坊的安全感，而且还鼓励了其他人撑起店面开展商业活动。

_Description

Florence, this bank is an old factory building that was last used as a supermarket. This project involved developing a roughly 400-square-meter interior into a space that resembles a large public square.
The waiting area and tellers are in the center of the square. The other various functions of the bank are distributed around the perimeter of the square. A small gallery space, for exhibitions of ceramic art, has been set aside near the entrance. A series of decorative Serralunga vases have been placed above the offices like still life. The outside of the bank is identified only by a logo.

BANK IN MONTELUPO FIORENTINO

Design Company_**Massimo Mariani studio**

　　弗罗伦萨的这家银行即将要落户的这座大厦是一家旧厂房,之前是一家超市。这个建筑项目是要将这个约400平方米的室内空间设计构建成一个类似于大正方形的场所。

　　等候席位区和出纳员的工作岗位规划在正方形的中心位置,其他的工作岗位则依次分布在正方形的边长上。主入口处已留出一部分空间,设计规划成一个小型画廊,用以展示陶瓷艺术品。一系列的塞拉伦加装饰花瓶被安放在办公室的上方空间,如同静物画一般。银行大厦外只有一个象征其身份的标志。

_Description

The first of our 7 projects with World Bank, the Tokyo Development Learning Center was named "Join Tokyo" and was constructed as the flagship center within the Global Development Learning Center program.

The design mandate was to combine the high-tech elements of the center, with a subtle reference to Japan. The designers' approach was to represent the technology element with video screens and very modern materials such as Corian and laser etched non-directional stainless steel.

The Japanese element was achieved through patterns, forms and lighting details, with references to the Zen rock gardens and shoji paper screen lighting effects. Functionally, the center is capable of holding up to 3 simultaneous multi-point, broadcast-quality video conference sessions in purpose built rooms that can be reconfigured and joined together for different sized sessions.

WORLD BANK

Design Company_**Orbit Design Co., Ltd.**

| Photographer_Basil Childers | Area_600m² | Location_Tokyo, Japan | **basic info** |

世界银行七个项目之一的东京发展学习中心，命名为"加入东京"，属于全球发展学习中心的规划之一，被构建为旗舰中心。

其设计理念是要把该中心的高科技元素与日本微妙的参照物相结合。设计师所采取的方法是以荧光屏和极具现代化的材料，如可丽耐和非线性光切不锈钢来代替技术元素。

通过借鉴禅宗的岩石庭院和障子塑料纸屏幕光照效果来设计图案、形式和照明细节，使这种日式元素得以成功实现。从功能上来说，该中心能够容纳三间同步多点、高音质视频会议对话的专用房间，它们能够重新配置，连接不同规模的会议。

Designer_Alexey Kuzmin

BANK OF MOSCOW

| basic info | Location_Moscow, Russia | Area_650m² | Photographer_Alexey Knyazev |

_Description

The office of Bank of Moscow is located in the heart of Russian capital on two floors of the building which was built in the 19th century and situated at the intersection of two historical streets. Atypical architecture, historicity and specific decorations of the building became a source of inspiration during the work with the first level of the office, where was planned to put the reception and two departments of the bank. In decoration were used luxury wood panels. The very center of the floor is decorated with stained glass in the shape of a diamond. The second level of the bank is located in the attic – this premise has never been used, so that was done without the historical touch. The design of this part of the bank was developed as a simple, comfortable and modern office with the large open space for the staff, with some training rooms and rooms for negotiations, which can easily be transformed in size.

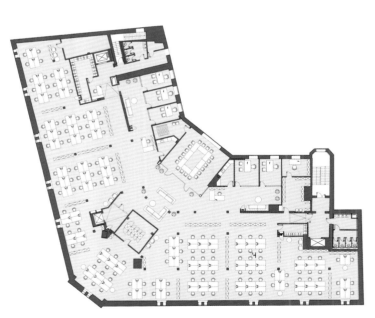

Mansard Plan

3 Floor

莫斯科银行的办公室占据了俄罗斯首都莫斯科中心区一座19世纪古建筑的两层，并且坐落于两个历史街区的路口。银行计划将第一层办公室设计为接待处和两个部门、非典型性建筑、历史性和建筑的具体装饰成为其灵感的源泉。装饰采用了奢华的实木面板。地板的中心由菱形彩色玻璃装饰而成。银行第二层办公室位于阁楼上，之前从未将银行设于阁楼中，这是银行史上的一次新尝试。这部分的设计包含能随意改变大小的训练室和谈判室，同时也为员工营造了一个简单、舒适、现代、开放、大面积的办公场所。

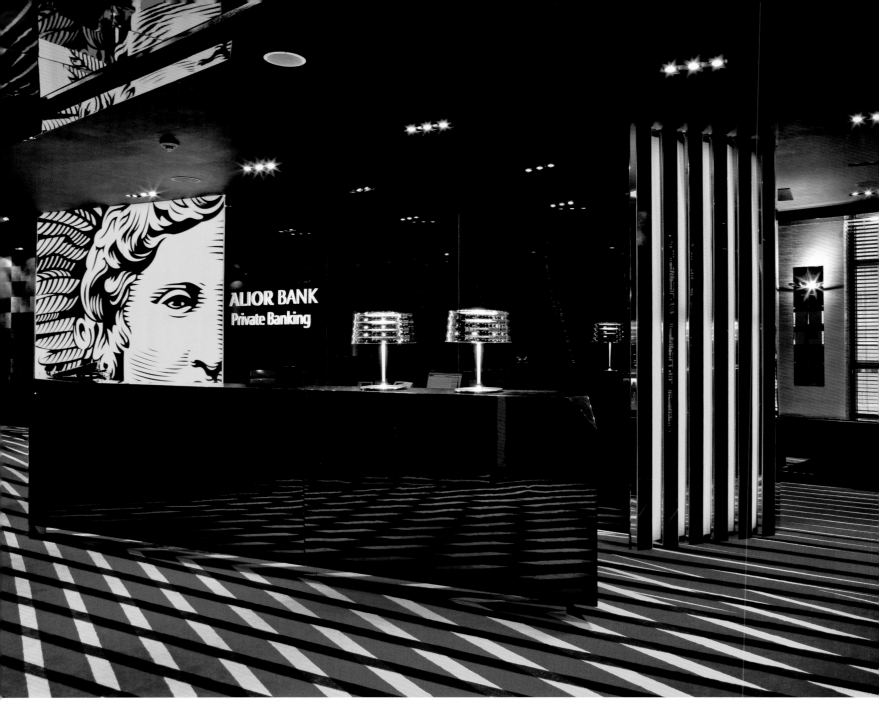

ALIOR BANK'S PRIVATE BANKING CENTER

Design Company_**Robert Majkut Design**

_Description

The main assumption of the project concerned the development of the very characteristic and strong bank's identification created by White Cat Studio.

The identification is based on three colors: crimson, yellow and white, and is styled as an old drawing – the face of an angel. This element is so characteristic that it became the starting point for further graphic works and studies.

The interior design concept of Private Banking outlets had to be supplemented. It also required designing the elements, which would differentiate it from the retail outlets. That's why black instead of crimson was incorporated into the sign. It creates a nice composition with other newly selected supplementary colors such as the shades of copper, grey and graphite.

The graphic linearity in the design appears not just in the rescaled logo dominating the reception zone, but is also included as the characteristic pattern on the carpet, multiplying the effect of perspective in space and referring to the effects observed in the sign.

Another aspect of the design transplanted from the visual identification concerns the shape consisting of two arranged squares. As a starting point they became the leading theme of all forms and shapes of the interior reminiscent on one side of the paintings of Piet Mondrian and on the other side of the already classic Bauhaus school.

The rhythm of lines and geometry of the appearing and overlaying squares create the internal divisions, starting from the main space plan and finishing on the furniture detail of the conference table top. The linearity and rhythm of the squares became the pretext not just for the development of the forms but also for the purely decorative effects.

| Photographer_OLO Studio | Location_94 Aleje Jerozolimskie St., Warsaw, Poland | Designer_Robert Majkutt | **basic info** |

  该项目的设计构思主要考虑因素是白猫工作室所开创的极具特色且强烈的银行身份象征。

  身份象征以深红色、黄色和白色三种颜色为基调,并且雕刻有古典绘画作品——天使的脸,这种象征元素充分彰显其特色,因此成为了后期平面图像工作和研究的出发点。

  设计师在对私人业务销售点的室内设计理念除了进行补充,还要在元素的设计上使其区别于普通服务中心。这就是为什么以黑色代替深红色嵌入到符号上去。它为其他新定的补充色创造了和谐的成分,比如,红铜色、灰色和墨色共同产生的阴影。

  其直线形的设计不仅出现在接待区的调整后的标识上,更作为特征图被融入到地毯上,同时增加了空间中的视觉效果和符号的直观效果。

  由视觉象征所诱发出来的设计的另一面则注重两个排列好的正方形组成的图形。作为出发点,这两种设计元素成为了室内各种各样装饰(包括其中一边蒙德里安图画作品和另一边先前经典的鲍豪斯建筑学院)的主旋律。

  两个相互凸显且重叠的正方形所显现出来的线条和几何图的旋律,作为源头从主空间规划到会议室每个设备的表面都进行了内部分割,正方形直线形主旋律的采用不仅仅是结构塑造的需要,更是简约装饰效果的要求。

NOBLE BANK

Design Company_Robert Majkut Design

_Description

Noble Bank specializes in private banking. Its services are aimed at very wealthy clients. Noble Bank combines the knowledge and professionalism of their personal advisors with its unique, tailor-made financial products. The inspiration behind the project was based on old English interior bank design and also photographs from the mid 1920s, which evoked the tradition and history of those days.

The reception area, conference room, and corridor wall facings and shining veneer plate, are in dark, chocolate colors, and the ceilings are made of upholstered dark brown textile panels. There are elements of cornices and columns finished with veneer that evoke prestige. Recessed lighting was used in the ceiling; however, individually designed table lamps were emphasized as well as a limestone floor.

There are four private offices designed for individual meetings with clients. Each room is in a different color but together they create a flowing unity defined by colors such as warm beige, cream, gold, and sky blue. All rooms are equipped with round black glossy tables and four armchairs covered with sky blue suede. On the walls there is wallpaper with a unique design. The same design is replicated on the glass wall divider between the reception area and the conference room and magnifies the harmony of the entire design.

This type of design in the Noble Bank starts a new trend, while returning to the retro style it brings itself to a new level with respect to modern design.

Robert Majkut designed all furniture and the emblems used on the wallpaper individually, exclusively for the purposes of this project.

This entire project recognizes people who admire high quality, professionalism, and aesthetic perfection. This is a place for clients who are expecting superior services in Poland that are no different from similar financial institutions in London, New York, Zurich, or Monaco. Noble Bank offers to its clients the same world-class treatment, quality, and comfort as the others mentioned above.

Photographer_OLO Studio | Location_multiple locations all over Poland | Designer_Robert Majkut | basic info

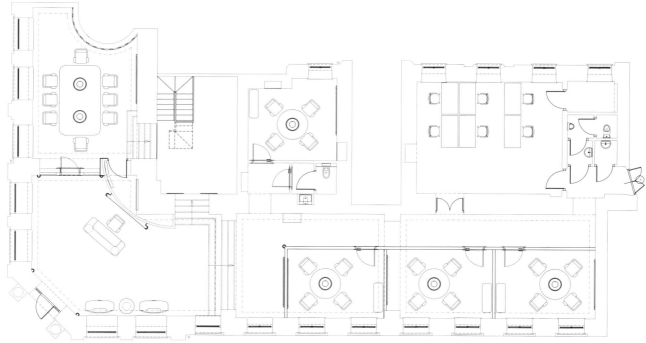

　　贵族银行专门从事私人银行业务。其旨在服务非常富有的客户。贵族银行凭借其独特的、量身定制的金融产品把知识和个人顾问的专业精神相结合。项目背后的灵感来源于老英国银行内饰和唤起20世纪20年代中期那些日子的传统和历史的照片。

　　接待区、会议室、走廊的墙壁饰面和闪亮的木皮板都是纯黑的巧克力色，天花板是由软垫深棕色纺织面板制成的。还有由单板所做的檐口和圆柱元素来体现其威望感。天花板中运用了镶嵌式照明灯、此外，单独设计的台灯也和石灰石地板一样被强调。

　　银行设有四个与客户单独会晤的私人办公室。每个房间都有不同的颜色，但它们一起创造出了一种由暖米色、奶白色、金黄色和天蓝色定义的流动一致性。所有客房都配有一张圆形的有着黑色光泽的桌子和四把天蓝色的皮毛扶手椅。墙上是设计独特的墙纸。接待区和会议室中间的玻璃分隔墙也采用了相同的设计，扩大了整体设计的和谐性。

　　贵族银行的这种设计引领了一种新潮流：复古风在遵循现代设计的同时把其上升至一个新的水平。

　　Robert Majkut专门为这个项目设计了所有家具和单独使用在壁纸上的标志。

　　这整个项目体现了人们所崇尚的高品质、专业性和审美的完美性。在波兰的客户能够享受到该银行提供的优质服务，其服务不亚于伦敦、纽约、苏黎世或摩纳哥等类似的金融机构。贵族银行为客户提供了和上述银行同样的世界级的服务、品质以及舒适性。

HSBC

Design Company_**Architect José Orrego**

_Description

Designing one of the world's most important Bank's headquarters in Lima required managing a contemporary and innovative look. Based on this, a perfect entrance was developed, so that corporative colors and sophisticated image would be shown.
Lighting is used as an enhacing resource: a red counter, a small fully equipped waiting area, and all inner rooms are absolutely special and comfortable spaces.
Social interaction and team work are very important for an optimal development of this office. Therefore, every space is designed and placed in a free plan to maximize perspectives and views. The cafeteria is conceived as a small and colorful "oasis" to take a break from daily work. Work stations use ergonomic furniture in adequate spaces designed for the HSBC dynamics.

basic info Designer_José Orrego – Metropolis Location_Lima

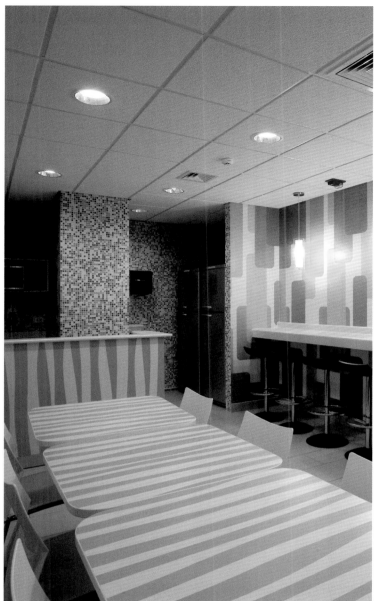

在利马设计一座世界上最重要的银行之一的总部，要求设计师成功打造具有当代意义，又很新颖的外观。在这个基础上，一个完美的入口就建成了，这样一来公司的颜色和精致的公司形象就展现在了人们面前。

灯光则被用做一种增强亮度的资源：红色的柜台，设备齐全的小型等候区，同时所有的内部房间都是相当别致和舒适的。

社会交往和团队合作对这个办公室的良性发展是非常重要的。因此，每块空间的设计和位置选择都是为了在自由的计划讨论中，大家可以各抒己见，发表自己的看法。自助餐厅被看成是一个小型的多姿多彩的舒服地，人们可以在日常工作间歇时，在这里小憩片刻。工作站宽敞的空间里摆放的是人体工程学的家具，这样的设计是为了与汇丰银行独具活力的特点相匹配。

CITIGOLD SELECT

Design Company_**Pentagram**

_Description

The all-encompassing design program for Citibank, the world's largest financial company, includes brand identity and interiors for Citibank branches worldwide. Whilst ensuring that the CitiGold "affluent customer" environments are consistent with the brand's core values, the palette employed clearly distinguishes the offer as an exclusive one, with clean, elegant lines, frosted glass partition and comfortable furnishing throughout.

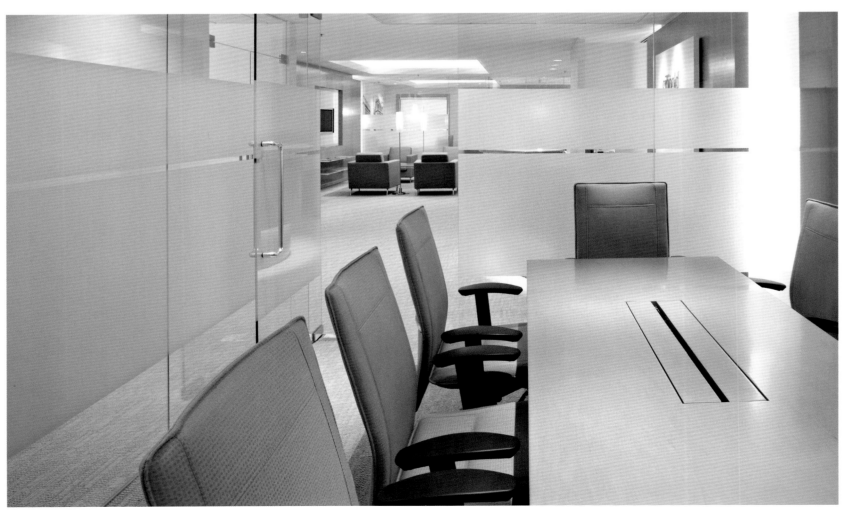

花旗银行是世界上最大的金融公司，本设计拥有全方位的设计程序，包括名牌识别和全世界花旗银行分行的内景。在确保花旗"富裕客户"环境与品牌核心价值观一致的前提下，简洁优美的线条、磨砂玻璃隔断、舒适的家具贯穿全场，此种设计基调使花旗顶级财富管理别具一格、独一无二。

INDEPENDENCE BANK HQ

Design Company_**AJArchitects**

basic info | Designer_Anton Janezich, Stan Andrulis, Shane Gerson, Katie Klos, Kun Zhang | Location_Washington DC

_Description

Independence Bank's goal was to ensure the financial needs of those not well served by the Capitol's existing banking institutions.

To retain a sense of continuity with the historic district's contributing buildings the design team retained and left exposed as much of existing interior brick bearing walls as possible. For historical sensitivity and cost the decision was made to gently clean the existing interior brick and then seal it with specialized polyurethane. The result was a synthesis of ancient walls and new furnishings that respected the integrity of the contributing walls while creating a whole appropriate for a modern banking institution. On the building exterior, dignified exterior signage and a new ATM installation appropriate to the epoch of the building was created. In a gesture to the overall historic district and time honored bank tradition, a public time piece, a cast iron pedestal clock was designed and is soon to be erected.

The institution's open plan with banking hall seamlessly integrated into the overall headquarters, creates a bright, crisp and friendly environment with the visual transparency Independence Bank sees for its future and the future of the US banking system as a whole.

The existing exposed interior masonry party walls are set off against 3.35-meter high white aluminum and clear glass doors, partitions and clerestories. The white interior finishes are illuminated with natural light provided by large renovated shop front windows skinned with glowing perforated shades. The resulting diffused natural light reduces the artificial lighting requirements during the day. In addition all aspects of the banking hall and offices are exposed to view for the staff bank officers and the general public to reinforce the sense of managerial banking transparency as well as allow natural light to filter through the space.

The bank set out to create a modest, light filled, open space that respects its context and is appropriate for the institution of the bank in today's world. It is clear these goals have been accomplished.

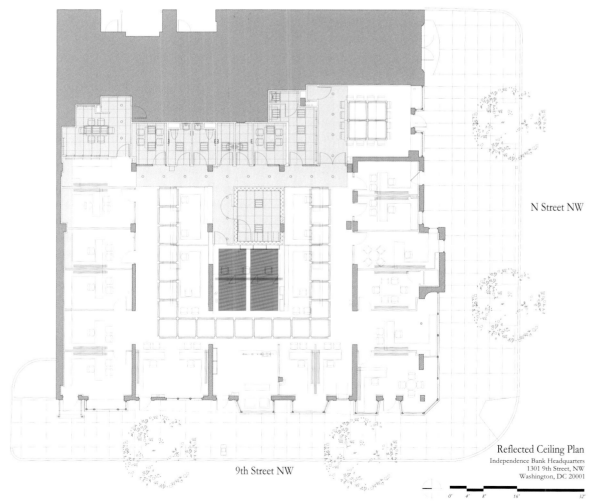

N Street NW

9th Street NW

Reflected Ceiling Plan
Independence Bank Headquarters
1301 9th Street, NW
Washington, DC 20001

0'   4'   8'   16'   32'

　　独立银行的目标是为了确保满足那些在国会大厦现有的银行业金融机构中未享受到周到服务的人的金融需要。

　　为了与历史区具有重要贡献意义的建筑保持一致，设计团队尽可能地保留了许多外露的现存的轴承墙砖。考虑到历史敏感性和成本，设计师决定轻轻地清理现存的内部砖块，然后用专门的聚氨基甲酸酯将它们封好。最终，它就是古代墙和新家具的完美结合，是它们的综合体，它既适合现代银行机构又与贡献墙相一致。在大楼外部有适合新时代建筑的庄严的引导标识和自动取款机。为了与整个历史区和由来已久的银行传统相呼应，设计师设计了公共时间作品和基座由铸铁制成的时钟，它们很快就会被竖立在那里。

　　该机构银行大厅的开敞布置与整个总部融为一体，与这个寻找自身未来和整个美国银行体系的未来的高度透明的独立银行一起创造了一种明亮、明确、友好的环境。

　　已经存在的外露的室内砖石界墙正好抵消3.35米高的白色的干净的铝合金玻璃门、隔墙和长廊。经过修复的较大的店面窗透过室内的自然光照亮了室内的白色家具，这些窗表面盖有发光的有孔百叶窗。这些自然光散布在室内，减少了白天人工照明灯的使用。此外，银行职员可以看见银行大厅和办公室的角角落落，同时公众可以更深层次地感受到银行管理的透明性，自然光则能充满整个空间。

　　该银行打算建造一个与其周边环境相一致的、适度的、充满阳光且开阔的空间环境，同时要适合当今的世界银行的设施。很显然，这些目标都已完成。

INDUSTRIAL AND COMMERCIAL BANK OF CHINA

Design Company_**JTCPL Designs**

_Description

Focusing its expertise, skill and energy, JTCPL Designs has designed a Bank in center city, Mumbai, seeking a simple, modern space that would immediately draw attention. ICBC appointed the architects to design the space that is notably open and inviting.

Industrial and Commercial Bank of China (ICBC), is amongst the foremost banking corporations of the world. With over 18,000 outlets including 106 overseas branches and agents globally, ICBC is one of the market leaders in commercial banking.

An office interior conveying the global image of the client was the prime requirement for the design team. Giving full consideration to the client's requirements, the design team meticulously sorted out the needs of potential customers, rationally dividing the space into multiple functional areas, maximizing the space utilization. Apart from being a race to the finish, the main challenge was to create a global office environment which would go with the client's worldwide presence as well as achieve impression of expanse in a conservative floor plate. The crux of the design is simplicity, functionality, brand image, and a blend of mild flavors from the Chinese and Indian cultures.

The space requirement was to accommodate a large banking business area, with a need for having a waiting area for client in close proximity of the banking area, therefore a lavish VIP Lounge was provided. The back office consisting of open office, conference room, GM's and DGM's cabin establishes a high level of design and finishes to affirm the company's importance. The office's spatial organization which started with a public zone at one end that becomes progressively more private at the other. This project promotes way-finding identity and intimacy of work area within the vast office space and creation of dynamic visual elements for staff and visitors. As a result, this expansive contemporary design is as inspiring as it is functional and efficient.

Area_ 697m² | Location_ Maharashtra, India | basic info

凭借其专业知识、技能和活力，JICPL设计公司已经在孟买中心城市设计了一所银行，目的是寻找一种可以立刻吸引人们注意力的简约而现代的空间。中国工商银行委任建筑师设计格外开阔而吸引人的空间。

中国工商银行是世界上最重要的银行之一。中国工商银行是商业银行的市场领袖之一，在全球有18000多家网点，包括106所海外分支机构和代理商。

对设计团队的主要要求是办公室内设计要传递客户的全球形象。充分考虑了客户的要求，该设计团队认真地整理出了潜在客户的需要，理性地将空间分成了多个功能区，将空间利用率最大化。除了要赶工完成建造外，设计师面临的主要挑战是创建一个全球化的办公环境，这样的环境既可以满足来自世界各地的客户需求，又能在一个保守的空间中给人一种宽阔的印象。这个设计的关键是简单性、功用性、品牌形象以及融合了中方文化和印度文化的文化氛围。

空间要求是必须要设置一个大的银行业务区，并且要在银行业务区附近安排一个大的客户等待区，因此这里有一个豪华的VIP休息室。这个由开放式办公室、会议室、总经理办公室和副总经理办公室组成的事务部门建立了高水平的设计成品部门。一端为公共区的办公室的空间结构在另一端却显得越来越私人化，员工有较大的私人空间。这个项目使得人们积极参与寻找方式的过程，同时在这个广阔的空间里，它也进一步增强了工作区的私密性，为员工和来访者创造了富有活力的视觉元素。结果，这个空间广阔的当代设计是鼓舞人心的，正如它功能性强，又有较高的效率一样，给人以信心。

JULIUS BAER PRIVATE BANK

Design Company_ **One Space Ltd**

| basic info | Project Leader_Greg Pearce | Location_Hong Kong, China | Area_567m² |

_Description

Julius Baer, a renowned Swiss private bank, engaged One Space to re-design their 567m² headquarters in Central Hong Kong. The brief called for a client reception area and meeting rooms, a private client dining room, a boardroom, and a highly transformable "Collaboration Lounge" accommodating ad hoc staff meetings and large scale get-togethers. The client requested "a bank that doesn't feel like a bank."

The successful outcome lies in the architects' candid and creative dialogue with the bank's CEO and COO, who brought superb vision and clear goal-setting. Northern European and Southeast Asian references and influences were called upon to coax out a unified atmosphere of cordiality, privilege and comfort.

A major design challenge at the "Collaboration Lounge" was to accommodate a wide range of activities in a single, transformable space. This room directly adjoins a 20-seat Boardroom. Sumptuously finished, extra-wide pivoting wall panels form the boundary between the two rooms, opening to 90-degrees to link the rooms and create a large multi-purpose event space holding up to 120 people seated and standing.

In addition, the client requested a "better-than-5-star" private dining experience for their invited guests within an intimate and distinguished setting. Concealed at one end of a unifying feature wall lies an exclusive Wine Room and Dining Room, supported by a professionally catered kitchen.

The architects optimized the use of sustainable materials wherever possible. Use of recycled or recyclable materials, such as aluminum timber and glass, form a large part of the interior finishes. Loose furniture and acid-washed river stone were sourced within an 805-kilometer radius of Hong Kong, reducing their carbon footprint.

Team Members_Winnie Cheng, Cosette Lai, Jan Chui, Selva Ku

Floor Plan

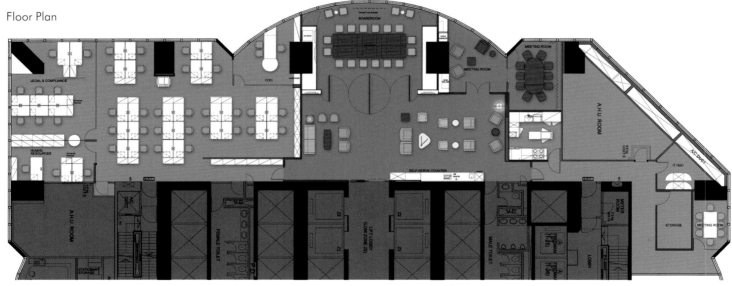

　　Julius Baer是一家著名的瑞士私人银行，它聘请One Space公司来重新设计其位于香港中环的面积为567平方米的银行总部。简单地说，它要求有一个客户接待区、几个会议室、一间私人客户餐厅、一间董事会室以及一间转换性极强的"合作休息室"，这间休息室可用来举行员工临时会议和举办大型的聚会。客户的要求是"建立一家不像银行的银行"。

　　成功的结果取决于设计师与银行CEO（执行总裁）和COO（营运总监）之间进行的坦诚且富有创造性的谈话，他们两位带来了极好的看法和明确的目标。为了成功地营造一种充满真诚、荣耀、给人舒适感的和谐统一的氛围，设计师还参考了一些北欧和东南亚的建筑。

　　建"合作休息室"时遇到的一个比较大的设计挑战是：它需要在一个单一的、可转换的空间内举办各类活动。该房间与设有20个座位的董事会会议室直接相邻。布置豪华且极宽的旋转墙板构成了两房间的分界线，旋转角度为90度，这样便将两个房间联系在了一起，同时还营造出一个可以举办各类活动的大空间、坐和站加起来最多可容纳120人。

　　此外，客户要求设立一间私人且别具一格的餐厅，那么他们所邀请的贵宾就可以享受"优于5星级酒店"的私人餐宴。掩藏在具有统一特色的墙体一端的是专门的饮酒室和餐厅，同时还带有一间提供专业美食的厨房。

　　无论什么地方，设计师都尽可能地充分利用可持续性材料。使用可回收材料，例如铝木材和铝合金玻璃，室内大部分成品都使用这些材料。宽松家具和酸洗的石头都是从半径不到805千米的香港本地购买的，减少了碳排放量。

ZÜRCHER KANTONALBANK

Design Company_**Retailpartners AG**

| basic info | Location_Kloten, Switzerland | Area_700m² |

_Description

The Zurich branches now greet their customers with a new "Look & Feel". Retailpartners AG made the brand values "personally, competently, responsibly" tangible and visible. The aim was to optimize the customer service at all points of interaction between the customers and the branch employees. The ambience in the renovated branches reflects ZKB's promise of quality; it is Swiss modern in functionality and quality, and unique with its atypical design for banks. The exquisite natural materials come from the Zurich region and underline the brand promise of the "nearby bank".

苏黎世银行的分行现在以全新的外观和风格来迎接其客户。Retailpartners AG建筑设计公司以设计出可见的有形实体将苏黎世银行"一对一私人化服务、称职优秀和负责任"的品牌价值观表达出来。其目的是通过改善银行员工和客户之间各个方面的互动达到优化苏黎世银行的客户服务的目的。整修重置后的分行大厦内的气氛布置反映和体现了苏黎世银行所承诺的高品质服务。银行大厦在功能和特性方面融入和体现了瑞士现代风格以及独一无二的不规则建筑设计。银行大厦所用的所有精美建筑材料都来自于苏黎世当地,强调了苏黎世银行所承诺的"家门口的银行"。

ICBC PERSONAL BANKING

Design Company_**JWDA**

basic info | Location_Shanghai, China

_Description

The private banking located in the Wealth Square of Pu Ming Road is the first Private Banking for ICBC which belongs to Shanghai Private Banking Headquarter of ICBC. It consists of two phases, 1,000m² for phase one and 1,200m² for phase two. Bearing many functions including lobby, multi-function section, training section, VIP section, conference room and administrative section, the private banking offers financial service for customers with over 8 million Yuan assets. Domestic design takes customer first principle as consideration and tries to create the image of high quality banking. Public sections are connected and open so that customers could feel unique sceneries of Huang Pu River the moment they come in. All the resources are combined in each activity that customer participates so that he or she could complete financial plans happily in an easy and elegant environment. In the multi-function section of Phase One, there is an open self-service section offering support for holding various types of parties. Phase two increases space for multi-function sector and adds small stage, food preparation room, audio control room and etc. This provides more flexible and open space for training, party and product exhibition. In designing private banking, traditional concept of banking is totally abandoned so as to make private banking a senior social occasion for elites.

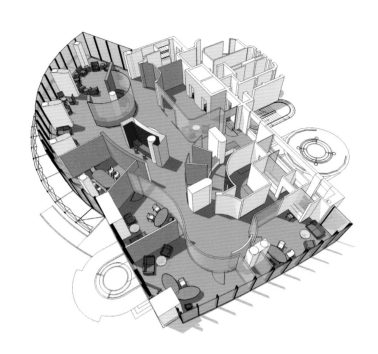
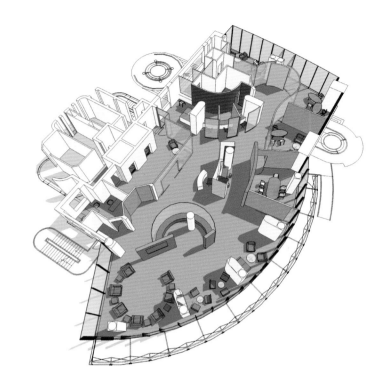

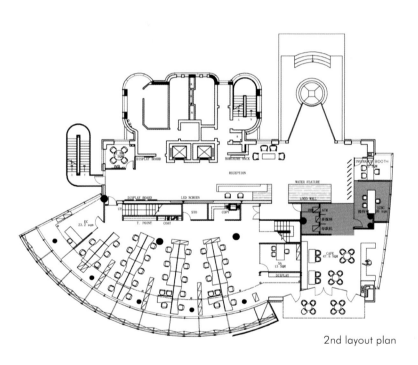

2nd layout plan

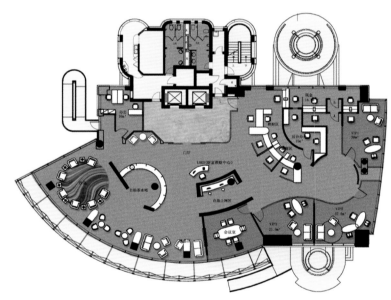

坐落于浦明路财富广场的私人银行是中国工商银行成立的第一家私人银行。属于中国工商银行上海私人银行总部。分为二期，一期面积约1000平方米，二期约1200平方米。功能分为门厅、多功能区、培训区、VIP接待室、会议室及办公区，服务的客户为800万元以上的理财客户。室内设计以客户至上为宗旨，精心打造一个高品质银行形象，使公共区域尽量连通、开敞，把有限空间无限延伸，让客户一进门便能感受到黄浦江景的独特资源，并把这些资源融合在客户的每一个活动中，使客户在轻松高雅的环境中愉快地完成每一次理财计划。一期的多功能区设置了开敞自助区，为举办各种宴会提供了辅助功能。二期增加了多功能区的实用面积，并设置了小型舞台、备餐间、音控室等。为各种培训、宴会、产品展示会提供了更加灵活和开敞的空间。在设计及服务流程上，颠覆了传统意义上的银行概念，使银行成为了精英们的高端社交场所。

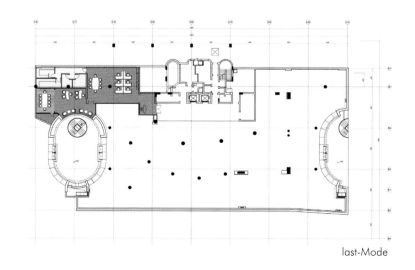

last-Mode

_Description

This corporation engaged the designers to analyze operations and develop a comprehensive Location Analysis to evaluate alternative locations in the Central business District for the 11,892 sqm occupancy. The bank's established space standards were respected, and merchandising criteria for building identity and branch legibility were taken into consideration. The selected location affords naming rights and prominent signage on the city skyline. Over seven floors, the bank's operations are arranged to co-locate interactive functions and maximize customer convenience. Support functions are clustered for efficiency. Existing furniture and equipment was reconditioned and redeployed with a vertical stack that places like materials from a diverse inventory on a given floor. The result is a cohesive whole, and achievement of the bank's objectives to minimize costs and re-use existing functional assets.

FIFTH THIRD BANK NORTHERN OHIO HEADQUARTERS

Designer_**William T. Eberhard AIA, IIDA**

Area 11,892m² Location Cleveland, USA

One-3rd floor plan

One-5th floor plan

这家公司要求设计师分析可操作性,并制定一个全方位的分析方案,分析这个占地11892平方米的中央商务区的位置。该银行确立的空间标准是受尊重的,同时考虑到了建筑特性和分行的易辨性的推销准则。所选定的位置给予冠名权并在城市风景线中着重标明。七层以上银行合作安排定位互动功能,这样最大限度地方便了客户。将辅助功能也都聚集在一起以便提高效率。现有的家具和设备都经休整并把其垂直堆叠,就像将材料从多样的库存中移至地板上一样。最后所呈现的是一个紧密的整体,最大限度地降低了成本,并重新使用了现有的实际功能。

One-6th floor plan

_Description

This project is intent to place a bank branch at the ground floor of an existing building in Ponsacco (Pisa). The building is on two levels: one out the ground, one underground, with a total surface area of about 400 square meters.
At the ground floor there is a big hall with four counters and the advisory services workstation behind which there is an archive and other administrative offices. At the same floor there is the branch directional bureau, a meeting room and two services offices. The underground floor houses technical rooms and few other offices. Vertical communications between floors is provided by a stair entirely coated in ceramic mosaic by Appiani.

BANK IN PONSACCO

Design Company_**Massimo Mariani studio**

Photographer Alessandro Ciampi   Area 400m²   Location Pisa, Italy   Designer Massimo Mariani

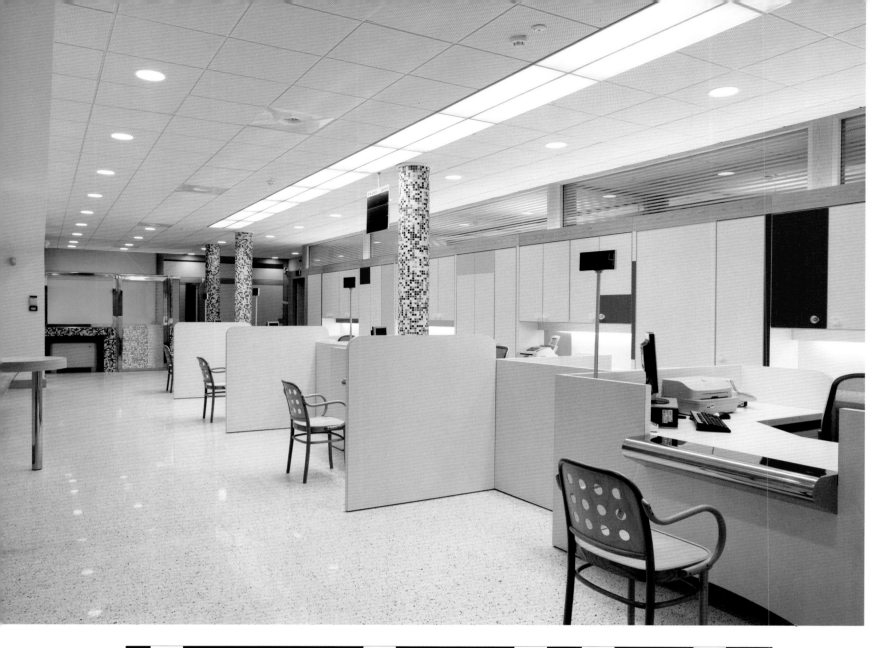

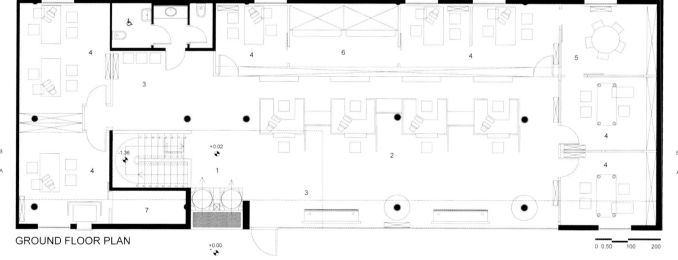
GROUND FLOOR PLAN

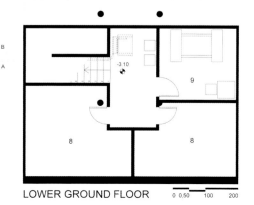
LOWER GROUND FLOOR

1. ENTRANCE
2. COUNTERS
3. WAITING AREA
4. OFFICE
5. MEETING ROOM
6. ARCHIVE
7. STORE ROOM
8. TECHNICAL ROOM
9. CAVEAU

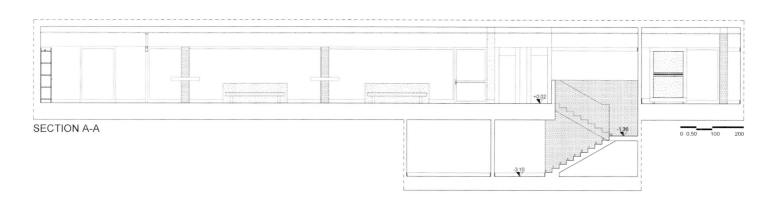

SECTION A-A

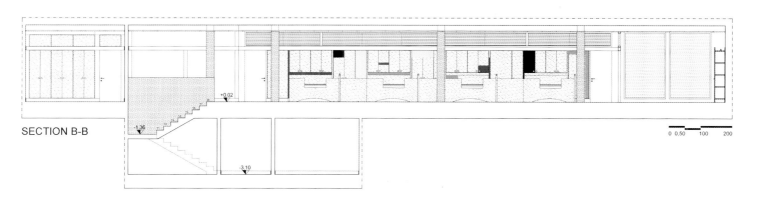

SECTION B-B

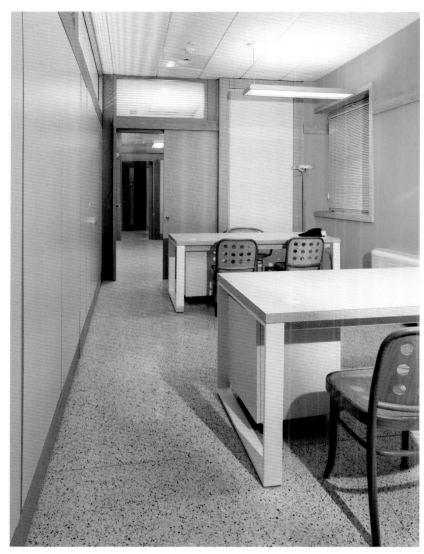

本案是要把一家银行的分行建在蓬萨科（比萨）一栋现有建筑的底层。该建筑共两层：地上一层，地下一层，总面积约400平方米。

地上一层有四个服务台和一个咨询站，后面还有一个档案室和其他行政办公室。在同一层，还设置了一个分行监管处，一个会议室和两个客服中心。地下一层则容纳了技术室和一些其他的办公室。楼与楼之间的纵向交流是通过Appiani所设计的楼梯实现的，其表面全部由陶瓷包覆。

GUARANTY TRUST BANK

Design Company_**Nicolas Tye Architects**

basic info   Location_London, UK   Photographer_Sarel Jansen

_Description

Located adjacent to the heart of London's west end shopping district parallel to Oxford Street, this busy central metropolis location has been historically know for its "rag-trade" industry. However, with ever increasing rental values and a relocation of various industries out of the area provided and opportunity for new retail and commercial based industries to start to establish themselves in this bustling location.
The acquisition of the entire building by Guaranty Trust Bank (GTB) in Margaret Street intended in establishing their European headquarters from this site. This location provides a high quality destination for its customers in a prestigious and highly accessible location, commensurate with the banks international reputation.
The building comprises of levels, ground, floors 1 to 5 and a substantial basement, all in covering some 1,858 square meters. The client (GTB) is a major banking institution based in Nigeria and provides banking services to net worth individuals, corporate trade finance and correspondent banking. It is the 5th largest bank in Nigeria and is highly respected in the international banking industry. It is the last few years winning some 7 internationally recognized awards. The bank maintains an extremely high credit rating reaffirmed by a triple A (AAA) risk rating for every year in the last 4 years.
The proposals were to create an executive boardroom level on the 5th upper most floor, providing the highest of quality to its finish and fittings. On the 4th floor there would be a series of waiting and meeting rooms along with a formal reception for high net worth individual clients facing the main street with the rear section of the building forming the banking representatives' office area. The remaining upper floors are fitted out to a very high office specification with Knoll bench desking, Vitra chairs, bespoke tea-points, Bisley storage units, Modular lighting and bespoke joinery. The ground floor would provide the retail banking facility with its astonishing stone finished open plan high specification layout, dramatic lighting providing a leading world-class banking destination. The basement level would contain the security vaults and safety deposit boxes along with a dramatic elongated approach corridor, reception and waiting facilities for high net worth individuals. The remaining basement area is being utilized for back of house training and storage.

信托担保银行邻近与牛津街平行的伦敦西区购物区，一个自古以来便以服装业而闻名的都市黄金位置。但是，随着日益攀升的租金价格以及政府提供的对各种搬离该地区的行业的重新安置政策，一些行业渐渐搬离，而一些以零售和商业为主的新行业则选择落户在这个繁华的地区。

被信托担保银行收购的位于玛格丽特大街的一整座大厦将被重新设计构建成其在欧洲的总部大厦。这个选址为其客户提供了一个交通便捷且知名度极高的地理位置，如同该银行在国际上所享有的声望一般。

该银行大厦有5层楼高，并且拥有一个大面积的地下室，总共占地面积约为1858平方米。信托担保银行是以金融机构为主，其总部设在尼日利亚，服务于高净值个人投资者、公司贸易融资以及代理银行业务。信托担保银行是尼日利亚的第五大银行，并且在国际银行业界内享有极高的声望。在这几年内获得了7项国际认可的大奖。在过去四年中，该银行在一个3A风险评级中都维持了极高的信用评级。

总部大厦的第5层楼的设计绝大多数是以董事会为设计主体的，并且提供最优质的设计构建以及相关的配件设施。大厦的第4层楼将设计构建一系列的会议室和休息等候室以及一个面向街道的、正式的招待处来迎接、招待高净值投资客户。招待处与同楼层后面的设计形成一个典型的办公区。剩下的楼层则按高级办公室规格来设计布局：Knoll长条板凳、Vitra椅子、量身定制的茶几、Bisley书柜、照明设施以及量身定制的木制工艺品。大厦的一楼大厅放置一些银行零售业务性质的机器设备，一楼室内用精美的石材饰面将其设计成开敞式、高规格的布局。变化莫测的灯光效果成功地将其打造成一个处于世界级领先水平的银行。地下室则设计成保险库和安全保险柜。此外，还有一条细长的走廊以及招待高净值投资客户的招待处和等待席位；其余的地下室空间则被规划设计成员工内部培训基地和存储室。

STANDARD BANK LONDON

Design Company_**Gensler**

| basic info | Location_London, UK | Area_11,148m² | Photographer_Owen Raggett |

_Description

Standard Bank's decision to relocate its 1200 London staff to new city offices began the process of the South African financial institutions strategy to provide a high quality work experience with enhanced brand visibility to the UK's financial community. 20 Gresham Street was chosen for its locality, being central to London's financial district.
Taking 4 floors over the ground, first, second and third floor, the Bank retained Gensler to provide full scope interior design services in order to portray, through its workplace, a truly modern, international organization that recognized its South African roots and heritage. Gensler created a progressive, clean lined and sharp design solution to reflect the company's core values – business driven, efficient, service orientated and hospitable – whilst positively expressing the companies brand values. Although a multi-tenant building, Standard Banks brand values are experienced as soon as you enter the lobby.
Flexible space with clear furniture layout occupies all standard office floors. Keen to promote collaborative workstyles Gensler incorporated a number of breakout areas on each floor and used the atrium area of the building to provide a shared space for all users groups to come together, ensuring a collegiate approach. Working closely with the South African client, careful attention was paid to referencing the Banks South African heritage, through the use of materials such as dark woods and crafted carpets the design aesthetic subtly alludes to this suggestion. State of the art "Bond Style" executive and client suites provide full conferencing and hospitality facilities with cutting edge AV technology enabling the Bank to cater for all client eventualities.

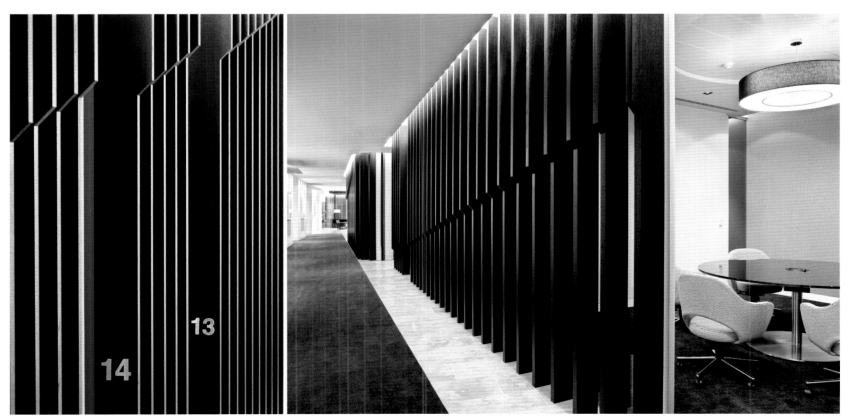

伦敦标准银行决定将其1200名伦敦员工迁往新城区办事处以便给英国金融界提供高品质的工作经验,提高其品牌的知名度,进而开始南非金融机构战略的进程。格雷欣20大街被选为其新址所在地,因为它是伦敦金融区的中心。

占地四层,地下室、一层、二层和三层。Gensler设计公司再次对其进行了全方位的室内设计服务,试图通过其办公场所来体现其现代化国际组织的身份,同时保持其南非传统及文化遗产。Gensler创造了一个渐进的、线条清晰的、利落的设计方案来反映该银行的核心价值观——业务驱动、高效、服务至上、热情好客,同时还积极地传达了银行的品牌价值。尽管是一幢多用户大楼,您仍然能一进大厅就感受到标准银行的品牌价值。

所有的标准办公楼层都有着灵活的空间,布置着简洁的家具。为了积极推动合作式工作方式,Gensler使每层楼都包含多个突围区,把大楼的中庭提供给所有用户群体共享,让他们聚集在一起,确保学院模式。由于与南非客户合作密切,因此该银行十分注重南非传统的营造,黑木和精心制作的地毯等材料都可以暗示这一点。现代化的"债券形式"和客户套餐使银行能迎合所有的客户,同时,银行可用最新的视听技术向顾客提供完整的会议技术和宜人的设施。

HSBC REGIONAL EXECUTIVE FLOOR

Design Company_**One Space Ltd**

basic info | Location_Hong Kong, China

_Description

In the first re-design of HSBC's Regional Executive Floor since this building was completed in 1985, the designers worked closely with the Executives to capture varying requirements and viewpoints, forging consensus from Concept to Completion. The dramatic facelift brings the regional executives together on one floor, facilitating more interactive engagement amongst the business leaders. VIP waiting areas and artwork galleries are opened to stunning views of the Harbor and cityscape for the first time in 23 years. The designers' strikingly modern re-design is a transformational departure from its predecessor, recognizing that, whilst built as the global headquarters of a Hong Kong bank, this is now the regional headquarters of a global bank. Significantly, the executive's own offices are largely transparent. An enigmatic "veil" embedded into the continuous floor-to-ceiling glass office fronts comprises two layers of crisply scored glass illuminated by concealed LED lighting, providing semi-obscurity and a luminous internal façade that adjusts with daylight.

The visual identity and operating functionality of this floor now underscore the vibrancy and emerging-markets-focus of this C-Suite to help strengthen regional leadership collaboration. Technology is fully integrated into every aspect of the interior, from customized financial broadcast video installations and state-of-the-art security systems, to high-definition videoconferencing incorporated into the wall of every office.

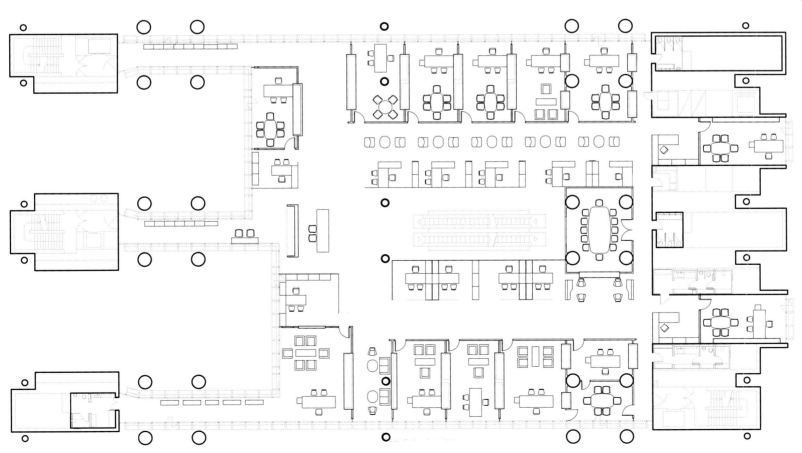

汇丰银行区域行政楼层始建于1985年。在首次重新设计时，设计师与银行高级管理人员密切合作，找出他们的各种要求和观点，并将其从理念转变为现实。通过大刀阔斧的改变，汇丰的区域高管们变为在同一楼层工作，这增加了他们与商业领袖间的互动。贵宾等候区和画廊在23年间首次面向港口和街景开放。设计师的现代设计和其前任的不同之处在于把一个地区银行的全球总部变为了一家国际银行的区域总部。高管的办公室大多透明，如面纱般的磨砂层嵌在接待台的落地玻璃上。两层金色玻璃、在隐蔽LED灯的照射下显示出朦胧美，且在日光下熠熠发光。

楼层视觉识别和运营功能凸显出活力和汇丰高管对新兴市场的关注，并借此强化区域领导力。从定制的金融视频播放设备、先进的安全系统到每间办公室墙上的高清视频会议系统，技术完全融入到了楼层内部的方方面面。

KOREAN EXCHANGE BANK

Design Company_**idpm Pty Limited**

basic info Area_330m² Photographer_Tyrone Branigan

_Description

This is not the first banking fit-out idpm has done in Chifley Tower, Sydney having also completed Sumitomo Mitsui Banking Corporation. They have had quite a lot of practice in how to make full use of the extraordinary views across the Botanical Gardens and Sydney Harbour right up to the heads, not to mention resolving some tricky issues to do with confined space and intrusive neighbouring buildings.

So, the designers are on familiar territory with the Sydney offices of the Korean Exchange Bank, including the opportunity for some subtle invocations of an Asian aesthetic. Essentially, the design consists of a simple division into two areas – public front-of-house and the back-of-house work area. The public area consists of reception and client waiting lounge, a meeting room, a boardroom and, interestingly, the CEO's office whose glazed wall connects it visually with the client waiting area. The first tactic was to frame the view – which involved editing out the ugly rear of a neighbouring building. Hence, on arrival the visitor faces the geometric planes of the reception desk and its backing wall which allows a slot window view into the CEO's office on one side and a frame for the view out to the Harbour on the other while hiding the neighbouring building. The unifying theme here is transparency – transparency to the outside view, but also internally with the extensive use of glazed walls and glimpsed views through to the CEO's office and into the main work area through a window to the side of reception, but masked from arriving visitors by a blade wall.

As a counterpoint to the orthogonal organisation of the space, idpm uses a large, circular rug and suite of round "Lily" chairs. Likewise, it has balanced the hard surfaces of glass and porcelain tiles with fabric, carpet, timber and colour, adding texture with the limestone reception desk and through the use of Jarrah burr veneer.

The overall effect is one of order and modernity, but warmed and softened by natural materials, a deft use of colour and the stunning visual connection with the natural world of the Botanical Gardens and the Harbour.

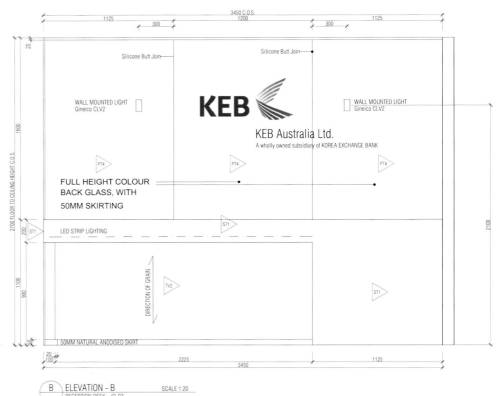

NOTE:
UNLESS OTHERWISE NOTED, ALL DOORS TO JOINERY TO BE BY HAFELE - CHROME PLATED, POLISHED DOOR HANDLE - 103.93.202 OR SIMILAR.
UNLESS OTHERWISE NOTED, ALL CARCASE TO TIMBER VENEER JOINERY TO HAVE BLACK MELAMINE AND ALL LAMINATE SURFACE JOINERY TO HAVE WHITE MELAMINE.

- TV1 — WOOD GRAIN TIMBER VENEER FINISH / CURLY BIRCH 60% GLOSS FINISH
- TV2 — WOOD GRAIN TIMBER VENEER FINISH / JARRAH BURR
- LM1 — LAMINATE FINISH / LAMINATE BY - LAMINEX / LAMINATE - CURLY BIRCH
- LM2 — LAMINATE FINISH / LAMINATE BY - LAMINEX / LAMINATE - PARCHMENT
- ST1 — STONE FINISH / POLISHED LIMESTONE
- PT4 — GLASS PAINTED FINISH / DULUX VIVID WHITE

B ELEVATION - B SCALE 1:20
RECEPTION DESK - JD-03

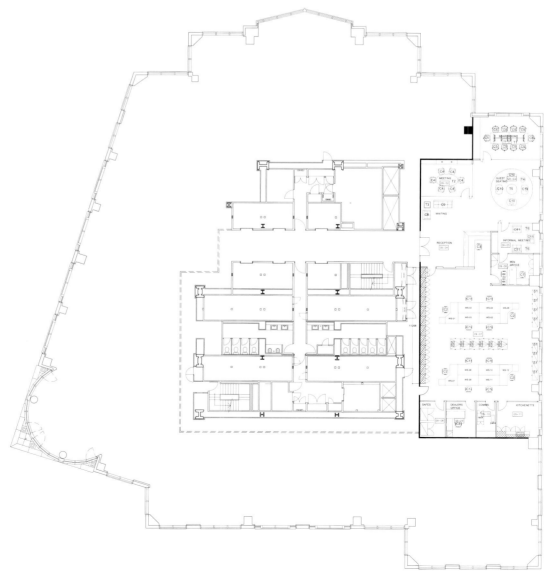

FURNITURE SCHEDULE			
ITEM	SUPPLIER	DESCRIPTION	QTY
C1	CHAIRBIZ	TASK SEATING - 'INCA'	8
C2	CHAIRBIZ	MANAGER 'BOXTA'	5
C3	CHAIRBIZ	RECEPTIONIST - 'BOXTA' POLISHED BASE	1
C4	BEVISCO	MEETING ROOM 'ULTIMA' WHITE	6
C5	BEVISCO	BOARDROOM 'ULTIMA' BLACK	9
C6	CHAIRBIZ	BOARDROOM 'LUXY' HIGH BACK - BLACK	1
C7	UCI	MD OFFICE 'CONTESSA' BLACK LTH SEAT	1
C8	CHAIRBIZ	WAITING AREA 'HEX' SINGLE LOUNGE	1
C9	CHAIRBIZ	WAITING AREA 'HEX' DOUBLE LOUNGE	1
C10	FY2K	GUEST SEATING - LILI	4
C11	FY2K	INFORMAL MEETING - FOSTER 500	3
T1	STYLECRAFT	BOARDROOM TABLE - CARMA BASE	1
T2	STYLECRAFT	MEETING TABLE - CARMA BASE	1
T3	CHAIRBIZ	WAITING AREA - HEX GLASS SIDE TABLE	1
T4	STYLECRAFT	GUEST & INFORMAL - CS GLASS SIDE TABLE	3
T5	STYLECRAFT	GUEST SEATING - CS GLASS COFFEE TABLE	1
T6	FY2K	INFORMAL MEETING - SIDE TABLES	2
S1	DEXION	2 TIER TAMBOUR DOOR STORAGE	8
S2	DEXION	3 TIER TAMBOUR DOOR STORAGE WITH TOPS	12

这不是idpm在悉尼Chifley Tower所做的第一家银行，之前他们还完成了Sumitomo Mitsui银行项目。该公司在如何充分利用植物园和悉尼港景观上做了大量工作，同时他们也很好地解决了对局促的空间和周边侵入建筑处理方面的一些棘手问题。

因此，设计师了解韩国外换银行悉尼分行的场地，也知道亚洲美学的一些内在要求。设计基本分为两部分：公共前台区域和后台办公区域。公共区包括接待、客户等候区、会议室、董事会会议室。有趣的是，通过玻璃隔板可以从首席执行官办公室看到客户等候区。首要策略是景观设计，包括去除大厦后面相邻的不雅观建筑。访客光临时可以看到接待桌上几何图案的飞机，透过桌后墙面一侧的窄缝可以看到首席执行官办公室，从另一侧缝隙可以看到港口，临近建筑则被墙面挡住。

此地装修的总体主题是对外界的景观透明。但室内广泛使用的玻璃隔板、背景墙把访客和首席执行官办公室隔开，通过隔板可以从办公室瞥见办公区接待处的景象。

为跟长方形场地相对应，idpm使用了大型圆地毯和系列百合花手扶椅。同时，玻璃面和瓷片在织物、地毯、木材和色彩的作用下和谐共处，石灰岩接待桌和红柳桉树饰面质感舒适。

建筑的整体效果是有序和现代。通过使用温暖和柔软的天然材料，熟练使用颜色，idpm把银行办公室和植物园及港口的自然景观完美地结合在了一起。

INDEX
索引

Alexey Kuzmin

Alexey Kuzmin, designer with fifteen years of experience, received the invitation to develop this design project from the Architectural bureau "Sretenka". Alexey is always trying to combine in his projects a pragmatic approach and aesthetics: " Design of interior is inconceivable without a clear sense of priorities. Functionality is a very important thing; each designer should pay particular attention to the people in their everyday life and holidays, worries and joys, their every morning, day and night". The office of Bank of Moscow was designed according to these principles, reflecting the bank's policy and corporate culture – respectable with the sense of reliability and trust, and thus convenient for its guests and employees.

Anton Janezich

Anton Janezich has spent nearly three decades working with some of the world's leading architectural firms and architects. His experience in designing and managing major projects in Europe, North America, Africa and Asia bring focused ability and tacit knowledge to every project he is involved in. Anton's architectural vision is driven by an awareness and application for the cultural, economic, and intellectual complexities of the modern world.

His experience includes modern and historic large-scale master planning, their component buildings and infrastructure, as well as more modest-sized corporate and residential projects, and collaborations in both building and interior architecture. Prior to becoming one of the founding architects of AJArchitects, Anton spent 18 years with Skidmore, Owings and Merrill and was an Associate Partner at their London office. He was mentored by associates of Mies van der Rohe's office, where he learned the design of exterior curtain walls from the experts who trail blazed this art form in the mid-20th century.
www.AJArchitect.com

Chris Rumble

Principal of IDPM Pty Limited, Master Project Director, Chris Rumble is a registered interior designer with the Design Institute of Australia and a Master Project Director (MPD) with the Australian Institute of Project Management. Chris has been Principal of IDPM Pty Limited since 2002 and led IDPM to receive two prestigious industry awards for Sumitomo Mitsui Banking Corporation and AXA Rosenberg. Chris has been involved in the design and construction industry since 1989 and has considerable knowledge of office banking fitouts having played key roles in many major fitouts in Sydney, Australia and the London, UK. Chris comes highly recommended by past clients as he is dedicated to producing outstanding results. His wealth of knowledge, diverse contacts, dedication, active project involvement and attention to detail guarantee successful outcomes for his clients.

Crea International

Crea International is a multi-disciplinary design company, a Thinking Lab that executes "Sense making" projects for branded architectural spaces, and whose Visual and Styling work is supported by comprehensive strategic planning and research.
Working through the unique method of physical brand design, Crea International's strength lies in integrating its extensive knowledge and experience in the delivery of strategically relevant and commercially successful brand experiences for its clients, by bringing the spirit and philosophy of the brand "to life" within the space, expressing its true soul. Their most prominent area of work is the retail sector, with particular strength in the banking & financial services sector and the telecommunications sector. However they work across an array of sectors including land & air transport, hospitality and food, fashion & beauty, entertainment, workplace and fast branding.

David Dalziel

David Dalziel co-founded the London design consultancy Dalziel and Pow in 1983. In his capacity as Creative Director, David specializes in a broad range of design issues, from environments to communications. He is instrumental in overseeing the successful interpretation of a client's brief and its transition, from concept to reality.
Dalziel and Pow have built a reputation for excellence in design, combing creativity and operational knowledge to deliver consistently successful projects in their clients. Dalziel and Pow are the UK's premier Retail Design Consultancy and are increasingly active international, with recent successes in new markets in India, USA, the Middle East and Asia.

DGJ

Founded in 1999, Drexler Guinand Jauslin Architekten is an international architectural office based in Zurich, Frankfurt and Rotterdam. In parallel to planning and realization, the firm research in the field of design methodology and sustainability. The integration of new technologies associated with a sustainable approach leads to creative projects, spanning from urban planning to interior design.
www.dgj.ch

Ellen W.H. Sander

Personalia: Ir. Ellen W.H. Sander, 28.01.1960 Warmond, the Netherlands
Education: Architecture, Technical University of Delft, 1981-1988
Final project: design of a "Water airport", situated at the "IJ" in Amsterdam.
Sander Architecten: Founding of Sander Architecten BV, November 1990 Haarlemmer Houttuinen 15, 1013 GL Amsterdam
Current Projects: Interior architect and supervisor New Office Rabobank Nederland in Utrecht.
Client: Rabobank Nederland. Sander Architecten is selected to create and supervise the execution of the entire interior design (56,000 m²), including the twentyfive-storey building. As the office interior is being redefined by the introduction of new methods of working, interior architecture is facing new challenges.

James White

James has been a member of the allen international team since 2003.
A Retail & Brand designer with over 12 years professional experience James has a wealth of experience working within multi-disciplinary design agencies for global retail and leisure brands including Vodafone, Heron City and Nike.
Before joining allen international James worked within the Brand Experience team at Enterprise IG (The Brand Union).
His experience also includes time at Fitch ,BDG McColl and Crabtree Hall working on a number of retail, leisure and brand focused schemes. These have included collaborating on the global retail guidelines for the launch of the Vodafone Live 3G services. Developing retail store concepts and masterplanning for the likes of Lloyds pharmacy, Alliance and Leicester and Foyles Bookshop.
James has a BA (Hons) in Interior Architecture and design from the Nottingham Trent University.

Gensler
Gensler

Gensler is a global design, planning, and strategic consulting firm networked across 35 locations on five continents. Consistently ranked by U.S. and international industry surveys as the leading architecture and interior design firm, Gensler leverages its deep resources and diverse expertise to develop design solutions for industries across the globe. For its longstanding commitment to the advancement of sustainable design, Gensler received the Leadership Award from the U.S. Green Building Council in 2005.

José Orrego

Architect graduated from the Ricardo Palma University Lima Peru (1987)
Member of the Architects Association since 1988.
Architect Orrego has specialized in different project areas where recent projects standout such as beach houses, corporate offices, commercial projects and large scale projects. Architect Orrego is also co-founder of the Peru Green Building Council.
In 1990 he won the contest for the recuperation of the historic city center "Lima can be changed" together with other distinguished architects.
In 1994 he was awarded the medallion of the city from the Miraflores district for the design of the Central Park that meant a significant contribution to the community.
In the 2005 he was awarded the PADIS prize for the commercial design category.
In 2006 he received two awards from the "XII Bienal de Arquitectura" with the projects HSBC offices and Pacifico Building.
In 2007 he was awarded the IIDA 2007 in the 34th annual edition IESNA LIGHTING DESIGN AWARDS where worldwide lighting projects are showcased. In this opportunity the PAUL WATERBURY FOR OUTDOOR AWARD OF EXCELLENCE was presented for the design of the corporate building PROFUTURO for the esthetic solution in the use of L.E.D.S
His work has been permanently published in different national and international magazines.
With more than 20 years dedicated exclusively to design he currently leads one of the most important architecture firm in Peru.
www.metropolisperu.com
info@metropolisperu.com

id.TECTURE

Hartono Dharmadi, director, with more than 20 years of experience in construction engineering, he masters the know-how in this industry. Worked both at architecture and interior field he soon establishes a wide network of worker and supplier in both field. With quality & client satisfaction as the goal, he directs this company to compete with other major architecture-interior company.
Rinto Wiharjo, design Director, graduated from Tarumanagara University School of Architecture in 2003, instantly after graduating he worked at a local design & build interior company that focused on corporate interior. In 2007 he got a chance to work in Singapore for short period of time at Artrend Design Pte Ltd and DP Design. In Artrend Design he learned more about residential interior and in DP Design he was introduced to the magnificent world of hospitality design. With these experience, in late 2008 he came back to Indonesia to start his interior design practice with his partner.
idtecture-studio@yahoo.co.id

Joseph Wong Design

ARCHITECTURE | PLANNING | INTERIOR

Founded in California in 1977, JWDA is dedicated to excellence in urban design, planning, architecture, landscape architecture and interior design. Its expertise spans a full spectrum of professional design services in hospitality, commercial, residential, educational and recreational communities, as well as data centers and urban complexes. JWDA is persistently committed to research in all new advanced building technologies and modes of sustainability, striving to create a better environment through quality designs and innovative solutions.

JWDA began to be commissioned for China projects starting in 1993. In 1997 they opened a representative office in Shanghai. It thrived and established itself as JWDA Shanghai Regional Headquarters in 2005, with another branch in Shenzhen opening just two years after. With quality design services at the core of its mission statement, JWDA China office is driven to embark on national and international expansion to better serve clients across China and the Asia Pacific.

JTCPL Designs

JTCPL Designs is an ISO 9001:2008 certified studio, specializing in creating aesthetic workplaces and is today recognized as a design company with incessant creativity, discerning taste and for meticulous execution.

At JTCPL Designs, they combine exceptional minds and gifted hands to create spaces visualized and crafted to the minutest details. Their team is a diverse mix of Architects, Interior Designers, Engineers and Contractors. Bringing together an ideal mix of theoretical and practical knowledge, the team has the capability to conceptualize the most optimal design and the skill to execute it to perfection.

Regardless of the size and scope of the project undertaken, the company has an excellent track record of delivering on time and within the client budget. The company focuses primarily on corporate office spaces ensuring deeper understanding of all design nuances, quicker response times, and flawless execution of all tasks at hand. Their designs are not only characterized by simplicity, but also by distinct lines and elegant forms incorporating a sense of movement, flexibility as well as scalability.

Robarts Interiors and Architecture

Robarts Interiors and Architecture is an interior design and architecture firm with offices in Beijing, Shanghai and Hong Kong. Established in 1996, the company is built around a credentialed and diverse group of over 70 professionals, including interior designers, architects, project managers, engineers and materials experts. Their completed projects in China include: corporate interiors, architecture, education and healthcare projects. No matter the scale or scope of the project, they are inspired by the opportunity to create meaningful projects where design integrity is supported by quality construction.

Kuadra Studio

Kuadra was established in 2001 by Andrea Grottaroli and Roberto Operti. They are a young and multitalented firm providing architectural design (civic, commercial and residential), industrial design and furnishings, interior design (residential, commercial and exhibition stands). They also offer graphic design solutions (from an integrated image for organizations and businesses, to the development of company logos, brands and internet sites). In addition to their professional services for clients, they also participate in many architecture and design competitions, winning many commendations and prizes for their entries.

Lauren Rottet

Lauren Rottet is one of the most-celebrated interior architects today with an extraordinary record of awards, publications, lectures, juries and honors. Ms. Rottet recently became the only woman in history to be elevated to Fellow by both the American Institute of Architects and International Interior Design Assoiciation. She also holds the coveted titles of Interior Design Hall of Fame, Contract (formerly Interiors) Designer of the Year and Lifetime Appointee to the U.S. General Services Administration's National Register of Peer Professionals for Design Excellence. In 2006, Rottet was inducted as an Inaugural Member of the Women in Design Hall of Fame. She also served as Juror for the 2010 Cooper-Hewitt National Design Award. Ms. Rottet's furniture and product designs have earned her four gold medals for Best of NeoCon and a Chicago Athenaeum Award.

Mark Horton / Architecture

Mark Horton/Architecture begins each project with the understanding that the questions which should be asked are more important than knowing the presumed final answers. The solutions will come at the end of a true design process, but only if the correct questions are developed at the start.

Mark Horton/Architecture integration of design aesthetics into the functional solution of the program at hand is of utmost importance. In the end, this direction becomes the primary goal of a process, the firm is centered around of engaging in the construction of architecture. To aspire to something different would be to resign the process to the idea that construction alone would suffice and that the need to engage an architect on the project would not be required.

Founded in 1987, Mark Horton/Architecture is a design firm based in San Francisco, California. With a primary focus on architectural design, projects have also been undertaken which deal with planning, interiors, and design aspects related to the architectural process.

Mark Horton/Architecture is a licensed architectural firm in the States of California and New York.

Lukas Fischbacher

NAME   Lukas Fischbacher Dipl. Interior Designer FH

PROFESSION   Interior Designer

BORN   18.09.1974

NATIONALITY   Swiss

EDUCATION AND QUALIFICATION

1991-1996
Apprenticeship interior fitting drawer

1999
Further education in colour scheme, Professional School for Colouring, Basel

2000 – 2003
University for Design and Arts, Basel

Feb 2010 – Current
Team Leader Planning and Design, Retailpartners AG, Wetzikon
Interior designs, point of sale designs

FEB 2008- FEB 2010
Project leader/Interior Designer, Creative Circle GmbH, Bern
Exhibition stand designs, event designs, interior designs, bar designs

2006 – 2008
Project leader/Interior Designer, Design Team AG, Zürich
Interior designs

2005 – 2006
Project leader/Interior Designer, Matthias Herzog GmbH, Laufen
Apartment buildings, familiy houses, interior designs

2003 – 2004
Diverse assignements

Massimo Mariani studio

Massimo Mariani studio, active for 30 years, dealing with projects of different nature; from architecture, interiors, design, it has a deep experience in the design of work places, offices and banks.
Massimo Mariani is the core design of the study where architects, designers, graphics, plus a qualified group of external consultants cooperate to important projects; a group of people from professional experience and many different configurations which produces a kind of cultural activity that becomes innovative ideas. All this makes the studio a reliable creative partner with outstanding organizational skills and coordination of all types of design experience.

Medusaindustry

Medusaindustry was established in 2009 as a spin-off company of medusa group architectural office and Art Plantation Foundation. The goal of the company is to realize projects in the field of art, industrial design and multimedia. Within the two years of its existence, medusaindustry achieved several important successes, among others: winning the competition for design of new interior standard for ING Bank outlets in Poland or the first prize for Jazz Club Tratata Caffe interior Architecture of the Year 2009/2010 of Silesian Voivodeship competition. The company also cooperates with music festivals like Tauron New Music or Off Festival (one of the biggest music events in southern Poland) by providing these with custom designed spatial objects and by arranging fragments of festivals' space. Medusaindustry is also an initiator of "Klopsztanga - carpet beating frame reactivation" action bringing back to the public space an object which primary function was cleaning carpets.
www.medusaindustry.pl

NAU Architecture

NAU is an international, multi-disciplinary design firm, spanning the spectrum from architecture and interior design to exhibitions and interactive interfaces. As futurists creating both visual design and constructed projects, NAU melds the precision of experienced builders with the imagination and attention to detail required to create innovative exhibits, public events and architecture.
www.nau.coop

Nic Preece

Nic joined allen international in February 2006 and has worked on a number of key strategic projects. Nic has held senior positions in companies both in the UK and abroad, as well as running his own business for nine years. He brings expertise in retail, leisure and workspace design, with extensive creative direction experience in graphic design and brand creation.
Nic spent a number of years living and working in Southeast Asia and India whilst under contract to TID International Pte. He worked as senior creative on the design and development of the Singapore Houses of Parliament, the design of the retail environment, streetscaping and food courts for Sun Tec City, the headquarters for Hill Samuel Bank and numerous other retail and leisure projects across the region.
Nic has an honours degree in three dimensional design from Kingston University and a diploma for completion of foundation art and design studies at Epsom School of Art.

Nicolas Tye Architects

Nicolas Tye Architects is an architecture firm based in Bedfordshire, United Kingdom that has designed many different types of projects for varying clientele. Some of their projects include schools, regenerated public spaces, retail spaces, office buildings, and homes. This award-winning company has created a gem with their office space that not only displays their capabilities, but gives their employees a wonderful place to work. They are architects and designers interested in creating innovative, high quality and healthy environments for all people.

One Space Ltd

One Space is the only integrated design and technology provider in the region that focuses on the financial services industry. To date, they have completed some 60 projects for large retail banks, global commercial & investment banks, hedge funds and brokerage houses. The founding Directors of One Space, Greg Pearce and James Oliver are a licensed architect and a technology specialist from the investment banking sector. They forged a strong and complementary professional relationship when they recognized shortcomings inherent in the traditional procurement of projects: The highly desirable synergy between technology and architecture seems instead often mired in adversarial relationships as opportunities for interdisciplinary creativity are recognized too late or not at all.
By contrast, One Space adopts a collaborative approach that fully understands the needs and aspirations of the client to open the widest possible range of holistic solutions.

Orbit Design Co., Ltd

Orbit Design is a creative multi-disciplinary design company based in Bangkok, with offices in London and Singapore. The company was established in 1996 and since their incorporation they have successfully completed a wide variety of design and construction projects both in the Asia region and internationally.
Orbit design has developed a reputation for innovative design solutions that are both memorable and successful. Their ability to encompass the 3 main design disciplines of Architecture, Interiors and Communications allow them to deliver truly integrated projects that invariably enhance and build business and brands.
www.orbitdesignstudio.com

Owen Raggett

Owen Raggett is a specialist architectural and interiors photographer based in Asia. Having relocated from London after 10 years of working in the design industry there, he now divides his time between China and SE Asia, and works with design firms, developers and publishers in the region.

Richard Benson

Richard Benson is one of a team of experienced retail designers at allen international. He has had experience across a broad scope of retail projects having worked on a number of high profile strategic design programmes since joining the company in March 1999. Before joining allen international, Richard worked as a freelancer in senior design positions for a number of projects in a variety of sectors. Banking and petrol retail have featured high up on this list. At Shell Petroleum, Richard worked on the client side with an inhouse team to develop concepts for Shell's future forecourt retail sites which involved corporate architecture, site planning and brand positioning. Whilst at John Ryan International, Richard was responsible for developing specific furniture designs for a radical redesign of Barclays Bank high street branches. At Lloyd Northover Citigate, Richard worked with a consultancy team to develop design schemes for Exxon Petroleum's (Esso UK) petrol retail environment designed for international roll out.

Pentagram

Founded in 1972 by three graphic designers, an industrial designer and an architect, Pentagram was soon acknowledged as one of the world's leading design companies with a reputation for ideas, craft, professionalism and wit. Over three decades Pentagram has consistently produced work whose variety and quality have earned the admiration of the design industry itself, clients and the public at large. Pentagram's design skills extend across the full spectrum of graphics and identity, architecture, interiors, products and exhibitions. Their multi-disciplinary structure, with teams from different disciplines working alongside each other, engenders a culture of cross fertilization that adds value to all creative thinking. Furthermore, it allows consistency of thinking through all design applications, whether two or three-dimensional, with the lead partner always having collaborative access to his or her partners. The fusion of these skills has driven many projects from retail design to aircraft interiors, museum, gallery and exhibition design and corporate interiors.

Silverfox Studios

Silverfox Studios is an Interior Architectural Design business focusing on the needs of the Hospitality Industry. The company was set up in 2007 in Singapore by the two Partners and co-founders Patrick Waring and Susan Heng who split away from design giant Wilson & Associates where they had been Deputy Managing Director and Design Director respectively.
Both Susan and Patrick have been at the forefront of Hospitality design for many years and have been responsible for the design and implementation of many luxury International Hotel Projects from China, India, the Philippines, Singapore, Korea, South Africa and Mauritius, to the Middle East and Europe.
The firm has a staff strength of 30 with designers from over 10 different nationalities providing accredited professional Architects, Graphic artists, Artwork Consultants, Product designers, Furniture designers and specialists in the formulation of procurement documentation and line item budgeting.
www.silverfoxstudios.com.sg

William T. Eberhard

In his 36 years experience as Director of Design and Principal at two of the region's leading architectural, planning and interior design firms, Bill has developed an expertise in a wide range of commercial, educational, institutional and retail facilities.
Bill's commitment to design excellence is reflected in the fact that his projects have been honored with over 40 major design awards at national, regional and local levels and he has authored numerous articles on strategic facility planning, renovation design, law office design trends, retail design trends and designing to improve workplace productivity.
Both a registered architect and certified interior designer, his reputation for crafting space and a seamless integration of products and materials has resulted in his selection as a member of national design advisory councils for Steelcase, Herman Miller, Haworth, AllSteel, Prince Street Carpet, Shaw Carpet, DesignWeave Carpet and Deca Limited. Mr. Eberhard was selected for Crain's "40 Under 40", serves on the American Institute of Architects Ohio Interior Design Licensing Task Force, is a founding member and President of the Design Forum of Cleveland, has served on the Board of Trustees at Vocational Guidance Services for over 15 years and he participated in the 2006 class of Leadership Cleveland.

Robert Majkut

Robert Majkut is one of the most important Polish designers. His hard-earned brand and consistency in his approach to design makes him recognizable as a very popular designer and creator of unique places. Born in Szczecin, Poland in 1971, he graduated with honours from an art secondary school with furniture design as major. Having first studied architecture at Szczecin University of Technology, he then moved to the Academy of Fine Arts in Poznan. Finally he graduated from Cultural Studies at Poznan University.
Robert's work received many accolades – in 2002 he was honoured with the "Rising Star" award by the British Council. He was also nominated for Elle Style Awards 2007 in the "good design" category. In 2011 Polish marketing magazine Brief included him in the ranking of 50 most creative people in business.
He has been active in the field of popularising good design. Member of the programme council of New Culture Bec Zmiana Foundation; guest speaker at universities, judge at competitions, expert quoted in Polish and foreign press.

Zoevox Architects

Zoevox is an interior design and design agency specialised in temporary constructions (stands, exhibitions, events) and in point-of-sales fit out and equipment supply (stores, POS, corners). Their performances range from conception to turnkey delivery. Their job tells a basic story to solve a sophisticated problem, whether it is a matter of temporary or long-term architecture, of scenography, liveliness, interior, graphic or product design. They consider converging complex parameters to reach creatively for a clear and elementary concept. They blend high-profile know-how in material craftsmanship, in space building, in specialised building trades. Their ideas grow richer from the materials features and techniques. They constantly look for materials that will best fit in the concepts the client want to develop.
They help clients specify and materialize their ideas. They help them to settle in and live in, to draw up a symbol, to show forward, to make a significant impact, to sell, to profile themselves in the cultural field, to express themselves… They strive for offering the clients a uniting concept, a story to tell which will make the place or space meaningful.

ACKNOWLEDGEMENTS

We would like to thank everyone involved in the production of this book, especially all the artists, designers, architects and photographers for their kind permission to publish their works. We are also very grateful to many other people whose names do not appear on the credits but who provided assistance and support. We highly appreciate the contribution of images, ideas, and concepts and thank them for allowing their creativity to be shared with readers around the world.